Intelligent Photonics Systems

The text comprehensively discusses the implementation of sensor systems using artificial intelligence, quantum device and circuit with artificial intelligence, and biomedical implementation of artificial intelligence-based optoelectronic circuits. It covers important topics such as photonic communication, optical and photonics neural networks, neuromorphic computing, quantum optics, and artificial intelligence leap in optical sensors, including biophotonics.

This book:

- Provides insights about the advancement in optoelectronics devices using artificial intelligence.
- Discusses artificial intelligence–integrated photonic crystal fiber-based devices.
- Covers artificial intelligence–integrated quantum networks for 6G communication and artificial intelligence–enhanced quantum optical devices.
- Presents neuromorphic computing for photonic circuits and deep learning–enabled optical sensors.
- Showcases development in artificial intelligence for biophotonics applications, deep learning revolution in optical networks.

It is primarily written for senior undergraduate students, graduate students, and academic researchers in the fields of electrical engineering, electronics and communications engineering, and computer engineering.

Photonics and Optical Communication: Technologies and Applications
Manish Tiwari

The field of optics and photonics is widespread, ranging from classic areas such as photography, cameras, lenses, displays and lighting, and spectrometers to more recent developments such as biophotonics, microwave photonics, nanophotonics, nonlinear and ultrafast optics, optical communications, and networks. This book series presents state-of-the-art research that focuses on success, theories, research practices, applications, and challenges in emerging technologies in the field of photonics, and communication engineering. Aimed at senior undergraduate students, graduate students, academic researchers and professionals, the proposed series will focus on the challenges in emerging technologies in the field of photonics, and communication engineering. The series will cover important topics including integrated photonics, visible light communication, nonlinear fiber, computational photonics, design, modeling and simulation, biophotonics, and nanophotonics.

Intelligent Photonics Systems: Technology and Applications
Manish Tiwari, Ghanshyam Singh and Ankur Saharia

Intelligent Photonics Systems

Technology and Applications

Edited by
Manish Tiwari
Ghanshyam Singh
Ankur Saharia

CRC Press
Taylor & Francis Group
Boca Raton London New York

CRC Press is an imprint of the
Taylor & Francis Group, an **informa** business

Designed cover image: shutterstock

First edition published 2025
by CRC Press
2385 NW Executive Center Drive, Suite 320, Boca Raton FL 33431

and by CRC Press
4 Park Square, Milton Park, Abingdon, Oxon, OX14 4RN

CRC Press is an imprint of Taylor & Francis Group, LLC

ISBN: 978-1-032-93944-5 (hbk)
ISBN: 978-1-032-96877-3 (pbk)
ISBN: 978-1-003-59103-0 (ebk)

DOI: 10.1201/9781003591030

Typeset in Sabon
by SPi Technologies India Pvt Ltd (Straive)

Contents

Editors

Manish Tiwari received a PhD in electronics and communication (ECE) in the field of photonics from Malaviya National Institute of Technology (MNIT) Jaipur. Presently, he is Professor and Head of Department of ECE at Manipal University Jaipur. He has more than 25 years of professional experience in various colleges and universities of repute. He has published more than 100+ research papers in reputed journals and conferences, and book chapters. He has also authored and edited more than a dozen books. He has also supervised more than 30 projects sponsored by DST Rajasthan. He is principal investigator in a BRICS project on Quantum Satellite and Fiber Communication (QuSaF) along with research partners in University of KwaZulu-Natal (UKZN) Durban, PSUTI Samara, and University of Science and Technology of China (USTC) Shanghai. He has also served on panels of experts and been the editor for various workshops by Commission for Scientific and Technical Terminology (CSTT), Ministry of Education of India (MHRD), Government of India. He has been visiting as a researcher to City University, London, under the UKIERI project in microstructured optical fibers during 2010 and 2011 and Tsinghua University, Beijing, during 2016. Dr. Tiwari has presented talks in PIERS at National Technology University (NTU) Singapore, APMP at PolyU-Hong Kong, TJMW at KMUT-Bangkok and Kasetsart University-Bangkok, City University, London, and several UKIERI workshops. He is on the panel of Traveling Lecturer Program of reputed societies like Society of Photo-Optical Instrumentation Engineers (SPIE) (USA) and OSA (USA) under which he has delivered invited talks in USTC Shanghai, University of Electronic Science and Technology of China (UESTC) Chengdu, Tsinghua University Beijing, ITS Indonesia, and UKZN, Durban (SA) on photonics technologies and career development. He is a Senior Member of IEEE and IEEE Photonics Society, Life Fellow of the Optical Society of India (OSI-India), Senior Member of OSA, and Member of SPIE and Fellow of Institution of Electronics and Telecommunication Engineers (IETE) India. He has been a Technical Program Committee member of many IEEE conferences and reviewer of IEEE/Elsevier/OSA/Springer/T&F journals. His current research interest includes free space and optical fiber communication, photonics, nonlinear optics, optical sensors, specialty optical fibers, space division multiplexing, structured light, photonic crystal fiber for nonlinear applications, space division multiplexing and sensing and photonic switching.

Ghanshyam Singh received the BTech degree from NIT Silchar, India, and MTech and PhD degrees in electronics and communication engineering from MNIT Jaipur, India. Currently, he is a Full Professor with the department of ECE, MNIT Jaipur, India. He worked as Visiting Research Scholar/Professor at HW University Edinburgh, United Kingdom, University of Eastern Finland Joensuu, and Keio University Japan. He has published over 150 research articles in peer-reviewed journals/conference proceedings. He is a Senior Member of SPIE, OSA, and IEEE, and a Fellow of OSI and IETE. He is the recipient of the 2017 IEEE Distinguished Lecturer Award of the Photonics Society. He is Chair (Elect) for OSA TG on Optics in Digital Systems and Member for SPIE Fellowship Committee. He is also serving as Secretary for the IEEE Photonics Society Rajasthan Chapter (PHO36), faculty advisor to OSA and IEEE Sensor Council Student Chapters of MNIT Jaipur. In the past, he had joint research collaborations with the researchers from Keio University (Japan), University of Vienna (Austria), LNPU (Ukraine) and Cairo University (Egypt). He is presently collaborating on a multi-institutional research project with researchers from the UKZN (South Africa), PSUTI Samara (Russia), and the USTC, Shanghai (China) for a BRICS STI Framework Program (2020–2023). His current research interest includes micro- and nano-structured photonic devices for integrated photonics, signal processing in optical circuits.

Ankur Saharia is currently working as an Associate Professor at Manipal University Jaipur, India. He has completed his PhD and MTech from the MNIT Jaipur, India, in the domain of integrated optics. He has more than 50 publications in the field of integrated photonics, optical sensors, and quantum logic circuit and has many international collaborations for his research. He has been an academic for eight years at the Poornima Group of Colleges where he has been recognized several times for producing outstanding contributions. He has organized many conferences and workshops and has also received a couple of travel grants to attend the prestigious annual SLC (student leadership conference) organized by OSA in Washington, DC (United States), and the leadership conference in 2021 in the United States. Since the beginning of his tenure at Manipal University Jaipur, he has organized numerous international events, such as webinars, conferences, and workshops with the participation of several distinguished speakers and renowned scientists, as well as hundreds of scholars from all over the globe and has inspired them to work in the domain of photonics and optics.

Contributors

Sajal Agarwal
Department of Electrical and Electronics Engineering
Rajiv Gandhi Institute of Petroleum Technology
Jais, Amethi, Uttar Pradesh, India

N. Alshaer
Faculty of Engineering
Tanta University
Tanta, Egypt

Kamal Kishor Choure
Department of Department of Computational Intelligence
SRMIST
Chennai, Tamil Nadu, India

Shashank Kumar Dubey
Department of Electronics and Communication Engineering
Global Academy of Technology
Bengaluru, Karnataka, India

Noonepalle HariPriya
Department of Electronics and Communication Engineering
G. Pullaiah College of Engineering and Technology
Kurnool, Andhra Pradesh, India
and
Department of Electronics and Communication Engineering
Koneru Lakshmaiah Education Foundation
Vaddeswaram, Andhra Pradesh, India

T. Ismail
Professor in Telecommunication Engineering, Taibah University

Santosh Kumar
Professor, Department of ECE
KL Deemed to be University
Vaddeswaram, Andhra Pradesh, India
and
Shandong Key Laboratory of Optical Communication Science and Technology, School of Physics Science and Information Technology
Liaocheng University
Liaocheng, Shandong, China

Nitesh Mudgal
Department of Electronics and Communication Engineering
Poornima College of Engineering
Jaipur, Rajasthan, India

Yatagiri Telugu Murali
Department of Electronics and Communication Engineering
G. Pullaiah College of Engineering and Technology
Kurnool, Andhra Pradesh, India

Manisha Prajapat
Department of Electronics and Communication Engineering
MNIT Jaipur
Jaipur, Rajasthan, India

Ankur Saharia
Department of Electronics and Communication Engineering
Manipal University Jaipur
Jaipur, Rajasthan, India

Than Singh Saini
Department of Physics
National Institute of Technology Kurukshetra
Kurukshetra, Haryana, India

Ghanshyam Singh
Department of Electronics and Communication Engineering
MNIT Jaipur
Jaipur, Rajasthan, India

Gargi Srivastava
Department of Computer Science and Engineering
Rajiv Gandhi Institute of Petroleum Technology
Jais, Amethi, Uttar Pradesh, India

Sandip Swarnakar
Department of Electronics and Communication Engineering
Koneru Lakshmaiah Education Foundation
Vaddeswaram, Andhra Pradesh, India

Manish Tiwari
Department of Electronics and Communication Engineering
Manipal University Jaipur
Jaipur, Rajasthan, India

lifespan, are bulky, and inflexible, there is a need for a more sustainable solution in the age of AI and 5G. Wearable gadgets are garnering interest and being seen as a possible option due to their incorporation of modern technologies and potential for self-sustainability. The pursuit of sustainable power sources and self-powered electronics/photonics in wearable devices has garnered significant global research attention and effort in order to achieve superior output performance with optimal efficiency.

In recent times, the field of technology has witnessed the emergence of AI integrated with functional electronics, leading to the development of a vast range of intelligent systems. These systems are capable of detecting, analyzing, and making judgments using algorithms that are helped by ML. Furthermore, the exceptional performance of integrated photonics, such as its capacity to acquire sensory data at a fast rate, enables high-speed transmission that can meet the demands of big data analysis and advanced forms of AI. By integrating wearable electronics/photonics with AI technology and fast communication, wearable devices can conduct more intricate and thorough analysis on the collected training sets compared to traditional methods used in previous decades. The implementation of a practical personalized wearable health-care system necessitates not only a few sensors and displays but also a sophisticated sensor network. In 2019, Bao's group showcased a body area sensor network called bodyNET, which combines several chip-free and battery-free wearable sensors that are spread around the actual body. The sensor signals are wirelessly sent using passive radiofrequency identification technology (RFID) to the flexible silicon circuit that is affixed to a textile material. As the quantity of sensors placed in the bodyNet grows to gain a deeper understanding of system-level operations, the sensors will disrupt each other due to electromagnetic interference. It is necessary to enhance the communication bandwidth in order to promptly send all the sensing signals and get feedback from the cloud in the context of the Internet of Things (IoT). Wearable photonics shows potential as a viable option to enhance wearable electronics in complex systems. Utilizing the optical sensing channel, wearable photonic sensors are impervious to EMI and may effortlessly achieve gigahertz (GHz) data transfer rates, enabling real-time monitoring and feedback. In addition, mid-infrared (MIR) photonic sensors are compact devices that possess the potential to selectively detect substances without the need for labels, while also being nondestructive. MIR photonic sensors have extensive use in several fields such as environmental monitoring, health-care monitoring, and industrial inspection. Several photonic building blocks for MIR have been developed, which serve as the basis for the future implementation of MIR wearable photonics. Simultaneously, the photonic neural network is progressing as a supplementary technology to the current transistor-based electronic neural network. In the field of wearable photonics, research and development are focused on polymer-based flexible photonics due to their advantageous characteristics such as easy fabrication, high flexibility, and stretchability. Due to the fact that the majority of organic bonds have absorption properties that extend beyond the near-infrared region, which encompasses the electromagnetic spectrum from the approximate limit of human eye reaction to that of silicon, flexible photonics based on polymers are primarily utilized in the near-infrared and visible range. In 2014, a flexible tactile sensor array that directly responds to pressure was created using an early polymer waveguide (Figure 1.1). The emergence of a photonic wearable business is being facilitated by affordable, miniaturized optoelectronic technologies, self-sustaining power sources, and rapid signal transmission. This sector encompasses augmented and virtual reality, illuminated clothing, and health monitoring, among other applications.

Integrated photonics plays a vital and important role in these systems, particularly in flexible displays and wearable biosensors [4]. Researchers are currently shifting their focus toward the development of flexible organic light-emitting diodes (OLEDs) and their integration into displays made of flexible materials or textiles. Wearable optical sensors have the capability to monitor several vital signs, including heart rate. Additionally, optical chemical sensors can now noninvasively analyze the chemical compositions of body fluids. For instance, they can determine blood glucose levels by studying human sweat. The capabilities and complexity of heart-rate monitors will continue to advance, such as by monitoring oxygen saturation and arterial blood pressure. Wearables can also be used for health purposes such as detecting UV radiation and measuring pollutant levels, both indoors and outdoors. According to David Simpson, the director of business development for mobile sensing at Integrated Device Technology in San Jose, California, sensors with algorithms that utilize different optical wavelengths are employed to extract valuable data for precise measurements.

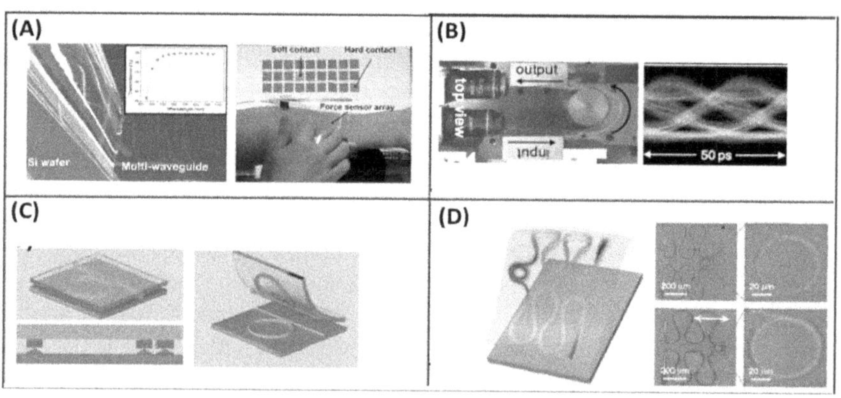

Figure 1.1 Showcases examples of wearable photonics: (a) A tactile sensor array based on polymer waveguides [1]. (b) High-speed data transfer utilizing multimode polymer waveguides [4]. (c) Silicon photonic circuits [5]. (d) waveguide-integrated photodetector [6].

Subsequently, information can be saved in nonvolatile memory devices embedded in the skin or sent wirelessly to external smart devices using an integrated Bluetooth module.

1.3 HOLOGRAPHIC NEURAL NETWORK (HNN)

AVR serves as an introductory AI problem at the start of a development route toward ASR [5]. An AI system that does not take into account complicated high-level cognitive processes is considered incapable of handling the entire ASR task due to its difficulty. ASR necessitates a profound comprehension of semantics, but AVR does not demand such extensive semantic understanding. An entry-level task refers to a task that has intentionally been made less mentally demanding, allowing for the exclusion of complex topics such as broad knowledge about the world. It does not simulate complex brain functions or maintain the belief that the brain is a crucial benchmark; hence, AI should demonstrate a significant level of plausibility in comparison to the brain. To ensure plausibility, it is necessary to establish fundamental principles of cognition in the brain. In brain-inspired AI, it is necessary to establish connections between low-level neural processes and high-level cognitive processes in order to attain plausibility. HNN adopts a contrasting bottom-up strategy, commencing with neurally plausible, low-level, distributed representations and thereafter constructing high-level cognitive conceptions. For the AVR task, just a small amount of high-level processing is required, making AVR an ideal starting place for HNN. Additionally, it includes a GPU. An essential concern in any AI strategy is the method of representing intricate information in a manner that can be effectively processed by the AI system. Hierarchies, tree structures, and similar networks of nodes can effectively model a significant portion of the information in the domain for most AI tasks. The task of AVR, as defined by [7], is a subtask within the broader field of ASR [8]. It requires a deep grasp of the meaning behind the speech in order to achieve performance comparable to that of a human. On the other hand, AVR is an introductory task that requires the capacity to generalize, but is less dependent on the surrounding context. This does not suggest that human vowel recognition is context independent. Instead, it indicates that the AVR task is deliberately designed to be independent of context in order to narrow down the scope of the work. More precisely, the text lacks specific details: Various individuals utter distinct single words, usually following a consonant-vowel-consonant pattern, such as "had" and "heed." The primary objective of the AVR task is to accurately categorize vowels produced by speakers who were not part of the training data based on a specific number of speakers used for training. Conventionally, ASR and AVR systems are perceived as separate components consisting of a feature extraction phase and a classification phase.

Feature extraction aims to decrease the dimensionality of the data while preserving the pertinent information. The methods employed for feature extraction include the following:

1. Mel-Frequency Cepstral Coefficients (MFCCs)
2. The Constant-Q Transform is a mathematical technique used to analyze signals in the frequency domain
3. Wavelet morphs

1.3.1 Classification

The majority of AVR systems employ supervised training. Various structuralist AI and ML techniques have been experimented with, such as hidden Markov models, support vector machines, and different ANNs. Recurrent systems have had limited influence in the AVR domain thus far. This could be due to the fact that repetition is not essential for the AVR goal since classifiers typically perform well even without recurrence.

In AVR systems, the packets are often evaluated using unseen data. However, in some instances, results have been documented using n-fold cross-validation, which means that the samples are not fully unseen if each speaker contributes more than one sample. A prevalent challenge faced by most classifiers is the high dimensionality of the feature vectors. In order to address this, it is necessary to reduce the input data into smaller feature vectors while retaining the same information. By prioritizing computational efficiency in the classifier, there is a trade-off of increased complexity in the feature extraction process and a certain degree of information loss. Some classifiers struggle with higher dimensional feature vectors due to reasons other than computational load, such as an elevated risk of overfitting. Establishing feature and dimensionality guidelines that are not solely based on convenience for a specific classifier is generally challenging. There is a limited availability of reference datasets for AVR. In many cases, the raw audio data is not accessible or the reference findings are focused on speaker identification based on vowels rather than directly on vowel recognition. We will briefly examine five sets of vowel data. The initial three datasets do not contain the original audio and are thus not immediately suitable for testing the simple HNN classifier. The last two datasets will be analyzed in relation to the previous three reference datasets.

The five datasets are as follows:

1. The Deterding's Vowel Dataset
2. The PBvowel Dataset
3. The Zahorian Vowel Dataset
4. The North Texas Vowel Database
5. The Hillenbrand Vowel Dataset

Vowel recognition was conducted automatically on two datasets: the North Texas vowel database and the Hillenbrand vowel dataset. The HNN classifier achieves comparable performance to existing effective automatic vowel identification methods, although a direct comparison with other classifiers has not been conducted. On one dataset, it can obtain a maximum of 74% accuracy in classification, compared to 84% accuracy achieved by people. On another dataset, it can get a maximum of 84% accuracy, whereas humans achieve 95% accuracy.

1.4 A PHOTONIC TENSOR CORE DESIGNED FOR MACHINE LEARNING

AI has achieved significant success in areas such as face and voice recognition, personalized music recommendations based on individual preferences, and the identification of cancers in medical images. The initial development of the ANNs occurred in 1950, but its widespread adoption has only taken place in the past several decades. A few years ago, Google recognized that the continuous advancement of AI could result in a higher energy consumption at its data centers. To address this issue, they created the Tensor Processing Unit (TPU), which saves energy through its universal programmability and emphasis on matrix operations. In the context of von Neumann architecture, matrix operations are executed sequentially [9]. However, specific architectures like graphic processing units and tensor processing units have been designed to conduct matrix operations in parallel [10]. It is important to note that these specialist architectures consume a significant amount of power. On a photonic platform, parallel processing can be performed efficiently and with minimal delay. This is made possible by the development of a photonic tensor core. It can reach a performance gain of two to three orders of magnitude compared to GPUs in computation.

A photonic tensor core is capable of performing 4 x 4 matrix multiplication [11] and accumulation without requiring additional electro-optic circuitry. In contrast, when matrix multiplication is performed on a regular central processing units (CPU), the operations are executed sequentially at each clock cycle, resulting in a complex computational algorithm. Therefore, the main objective is to reduce execution time and complexity in the computational algorithm. Integrated photonics is considered for this purpose, and photonic tensor cores are utilized to perform large-scale matrix operations. It is well known that digital electronics relies heavily on gates in integrated photonics. This means that multiplication, accumulation, and general linear algebraic operations can be carried out using the inherent parallelism offered by the electromagnetic properties of light [11]. The primary benefit of conducting matrix operations in the photonic domain is that they may be executed with minimal static power consumption and low latency.

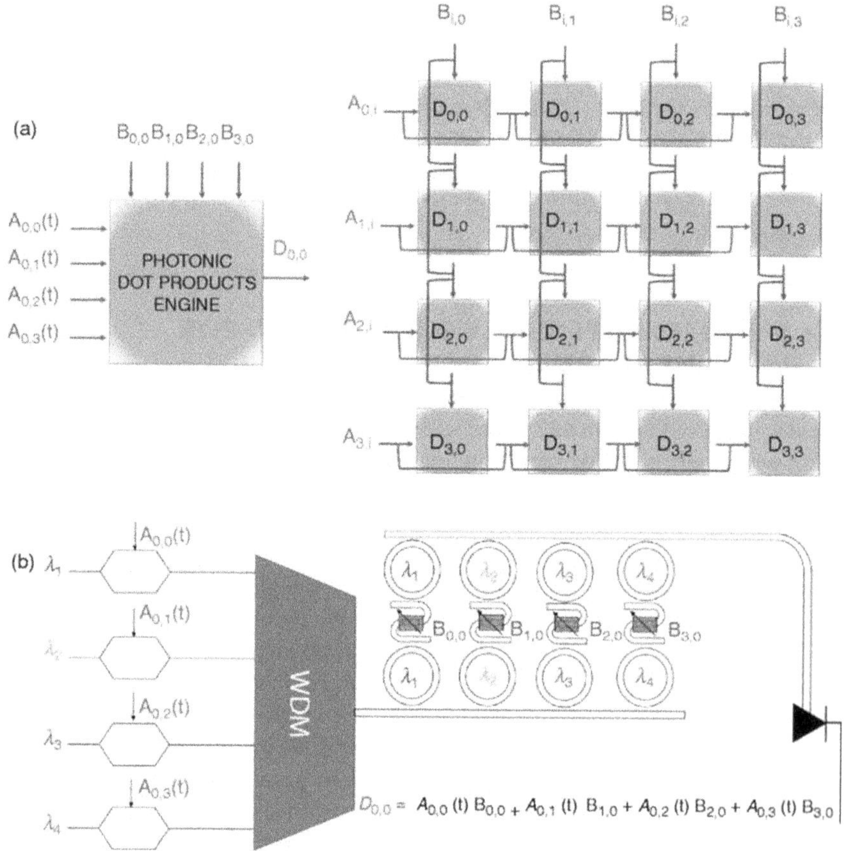

Figure 1.2 Photonic tensor core (PTC) [11]. and dot product engine. (a) A 16-dot product engine is a device that does the multiplication between two vectors. (b) The dot product engine is specifically designed to carry out the multiplication operation between two vectors.

A photonic tensor core utilizes 16 units known as dot product engines, which execute element-wise multiplication and incorporate Wavelength Division Multiplexing (WDM) technology, as depicted in Figure 1.2(a). The dot product engine, as shown in Figure. The operation 1.2(b) calculates the dot product between the ith row of the input matrix A and the jth column of the kernel B. The ith row of the input matrix consists of WDM signals, which are either in the optical domain or modulated by high-speed (e.g., Mach–Zehnder) modulators. The jth column of the kernel matrix is stored in the photonic memory by accurately configuring its weight states. The light-matter interaction is utilized to exploit the phase-change, material-based, photonic memory. The inputs are appropriately filtered by micro-ring resonators (MRR) to selectively extract specific spectrum components. These filtered inputs undergo a quantized electro absorption process, resulting in element-wise multiplication. The weighted inputs are summed using a photodetector, which performs a multiple accumulate computation operation, resulting in the element Dij of the output matrix.

1.5 OPTICAL NEURAL NETWORK

The researchers have constructed an ANNs model that mimics the functioning of the human brain, drawing inspiration from its physical neuron system. ANNs are capable of executing highly intricate logical and nonlinear processes, as well as distributing data in a parallel fashion. These significant capabilities have rendered ANNs crucial in the field of AI [12]. ANNs have achieved significant success in several ML tasks, including speech recognition, face recognition from photographs, medical diagnosis, and intelligent imaging. However, due to their increased computational demands, engineers are seeking alternative computing methods that are faster and consume less power. One of the most promising approaches is optical neural networking, which not only consumes less energy but also offers high efficiency, low crosstalk, and, most importantly, its intrinsic parallelism greatly enhances computing speed. An optical neural network is a physical realization of an ANNs that incorporates optical components. It has applications in image processing, pattern recognition, target tracking, and optical switching. Recent advancements in optical NN have achieved significant breakthroughs in executing highly intricate computations and design techniques through the utilization of cutting-edge intelligent design methodologies [13].

Optics or photonics are highly suitable for information processing due to its advantageous characteristics, such as minimal losses, low power consumption, wide bandwidth, rapid propagation speed, and crucially, absence of interferences. The photons can be merged based on factors such as time [14], polarization [15–16], momentum [17], and wavelength [18]. This method resolves the issues encountered when utilizing electronic equivalents. Hence, optical chips are currently demonstrating remarkable performance in the integration of circuits and sophisticated circuitry across various fields. The introduction of new advancements in information processing is always highly appreciated in all systems. The use of ANN-based systems introduced innovative advancements to optical domain technologies. An innovation based on intelligent systems termed reservoir computing was introduced in 2011. It is a technique used to effectively calculate complex network information. A proof of concept for an ANN utilizing optical fiber and based on reservoir computing has already been showcased [19]. Integrated photonics has the potential to be a superior choice for constructing circuits based on ANN. In the past decade, researchers have made significant advancements in the field of NN. In 2012, a team of researchers [20–21], successfully showcased a recurrent neural network that utilized an optical fiber technology. In this network [20], the researchers demonstrated the injection of an optical signal through an arbitrary generator, which can then be manipulated using modulators and amplifiers. The network includes intermediate levels known as reservoir layers, which are composed of optical delay lines, attenuators, detectors, and amplifiers. The system demonstrated intelligence by accurately detecting and identifying different signal waveforms through meticulous training and precise weight management.

1.6 NANOPHOTONICS AND DEEP NEURAL NETWORKS (DNN)

The topic of nanophotonics has had significant research activity in the last 20 years, driven by the growing fascination with investigating novel physics and technologies involving light at the nanoscale. Nanophotonics is the scientific field that investigates the behavior of light and its interactions with matter on a very small scale, namely at the nanoscale [22]. In recent decades, there has been a significant increase in interest in this sector, leading to active research and a wide range of applications. Traditionally, the design of nanophotonic devices relies on physics-inspired methodologies, even if they may have diverse underlying mechanisms, topologies, and materials. The early designs are typically evaluated by simulations that solve Maxwell's equations. However, it is unlikely that they will immediately achieve the necessary performance. Hence, it is necessary to repeatedly make modifications to a few parameters and reassess through simulations in order to come closer to the desired outcome. Deep learning is commonly employed to facilitate nanophotonics. Deep learning is a specialized area within the fields of ML and AI. Its remarkable achievements in computer vision and speech recognition, as well as its significant advancements in diverse applications like strategy development, have garnered growing interest. Nanophotonics is a distinctive field that leverages deep learning to design advanced devices and enhance the performance of existing techniques. Additionally, it enables the implementation of deep learning algorithms that can operate at the speed of light and with minimal energy consumption.

Nevertheless, after many years of ongoing enhancements in accordance with Moore's law, electronics began encountering limitations in these fundamental physical features that can no longer be overcome through scaling. Integrated photonic circuits are regarded as possible solutions to address the aforementioned challenges, because to the superior speed and energy efficiency associated with photons. However, conventional electronic components like CPUs are not suitable for supporting the new methodologies in AI.

Photonic circuits often run at higher speeds and with higher energy efficiency than electronic circuits. However, there are a many advantages to implementing DNNs and computing on photonic platforms [23]. The architecture of a DNN comprises numerous processing layers [24], with an input

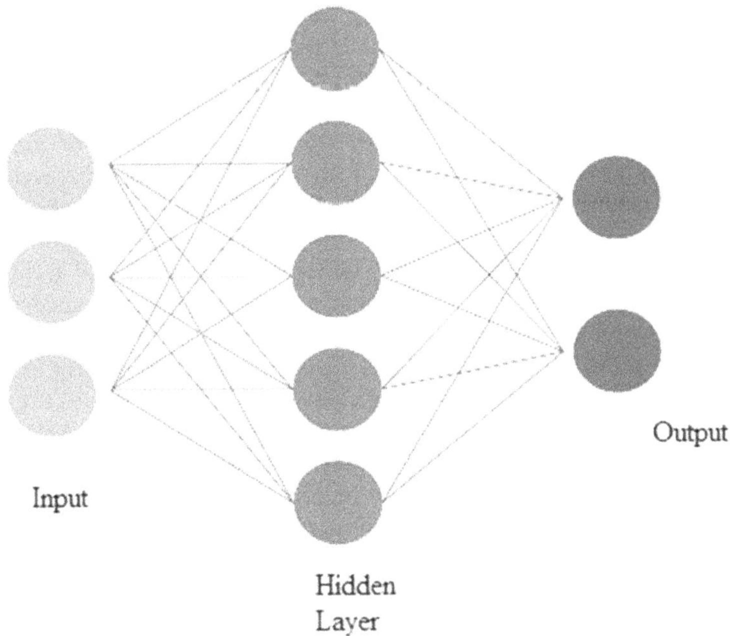

Figure 1.3 Structure of a DNN that consists of numerous layers.

layer at the bottom, an output layer at the top, and at least one hidden layer in between as shown in Figure 1.3.

The input and output layers of a DNN are comprised of a predetermined number of neurons or units. The size of the input layer is decided by the amount of data being fed into the DNN, while the size of the output layer is controlled by the specific task that the DNN is designed to perform. An early example of photonic DNNs was demonstrating vowel recognition [23], which attained performance comparable to a 64-bit electrical computer. In order to fully implement the map of a DNN after training, each layer of the network consists of an optical interference unit (OIU) responsible for performing the linear matrix multiplication, and an optical nonlinearity unit (ONU) that serves as the nonlinear activation function. The circuit was trained and tested using four distinct vowel phonemes spoken by a total of 90 individuals. Prior to being input into the nanophotonic DNN, the vowel phonemes were preprocessed on a computer and then converted into amplitude-encoded light pulses to produce outputs.

1.7 FUTURE SCOPE AND CONCLUSION

Photonic microchips, often known as photonic integrated circuits, are capable of doing all operations using light as the fundamental element. Photonics has significant developmental potential, but it requires comprehensive scientific research to effectively manipulate light for all tasks. Photonic microchips possess exceptional speed and can partially substitute microelectronic chips with great efficiency. Photonic microchips offer exceptional energy efficiency, allowing devices to operate for extended periods of time. The implementation of optical NN in industry is the upcoming breakthrough technology in this highly lucrative field. Currently, firms are directing their research toward neuromorphic and all-optical quantum computing as a means to develop future AI systems. When comparing optical technology to its electronic cousin, it is evident that optical technology is superior due to its low power consumption and latency. These qualities are crucial for AI-based systems. Companies such as Intel are making significant efforts to bring optical neural network-based systems to the market due to the highly promising situation. While ANN-based optical circuits are currently unable to rival electronic circuits, the future of AI computing may lie in the utilization of nanoscale photonic circuits. By incorporating three-dimensional optical integration and introducing an additional spatial degree of freedom, it would be possible to enhance the design of optical NN (ONN) by increasing the number of artificial neurons and enabling the creation of more intricate NN. In addition, the photonic sensor has the capability to identify biological neurons. The advancement in nanofabrication techniques is expected to significantly benefit the future of ONN. Essentially, the future of computing revolves around enhancing the immediate response times, energy usage, and ability to handle larger workloads of optical integrated NN.

REFERENCES

[1] Shi, Q, Dong, B, He, T, et al. *Progress in wearable electronics/photonics—Moving toward the era of artificial intelligence and internet of things*, Wiley 2020. https://doi.org/10.1002/inf2.12122

[2] Gao, W., Ota, H., Kiriya, D., Takei, K., & Javey, A. (2019). Flexible electronics toward wearable sensing. *Accounts of Chemical Research*, *52*(3), 523–533.

[3] Liu, Y., Pharr, M., & Salvatore, G. A. (2017). Lab-on-skin: a review of flexible and stretchable electronics for wearable health monitoring. *ACS Nano*, *11*(10), 9614–9635.

[4] Shi, F., Bamiedakis, N., Vasil'ev, P. P., Penty, R. V., White, I. H., & Chu, D. (2018). Flexible multimode polymer waveguide arrays for versatile high-speed short-reach communication links. *Journal of Lightwave Technology*, *36*(13), 2685–2693.

[5] Chen, Y., Li, H., and Li, M. Flexible and tunable silicon photonic circuits on plastic substrates. *Sci Rep* 2, 622 (2012). https://doi.org/10.1038/srep00622

[6] Li, L., Lin, H., Qiao, S. et al. Monolithically integrated stretchable photonics. *Light Sci Appl* 7, 17138 (2018). https://doi.org/10.1038/lsa.2017.138

[7] Kjeldsen, H. D. Automatic vowel recognition with GPU based holographic neural network. Diss. Faculty of Science and Engineering, 2008.

[8] Arora, Shipra J., and Rishi Pal Singh. Automatic speech recognition: a review. *International Journal of Computer Applications* 60(9) (2012).
[9] Coppersmith and Winograd, S. Matrix multiplication via arithmetic progressions. *J. Symbolic Comput.* 9(3), 251 (1990).
[10] Nicola Peserico, Bhavin J. Shastri, and Volker J. Sorger, Integrated photonic tensor processing unit for a matrix multiply: A review. *J. Lightwave Technol.* 41, 3704–3716 (2023).
[11] Miscuglio Mario, and Sorger Volker J. Photonic tensor cores for machine learning. *Appl. Phys. Rev.* 7 (3), 031404 (2020). https://doi.org/10.1063/5.0001942
[12] Goodfellow, Y. Bengio, and A. Courville, *Deep learning.* The MIT Press, Cambridge, MA, 2016.
[13] Saharia, Ankur, Choure, K., Mudgal, N. et al. Introductory review on all-optical machine learning leap in photonic integrated circuits. *Opt. Mem. Neural Networks* 31, 393–402 (2022). https://doi.org/10.3103/S1060992X22040075
[14] Deng, R. R. and Liu, X. G. Optical multiplexing: Tunable lifetime nanocrystals. *Nat. Photonics* 8, 10–12 (2014).
[15] Zijlstra, P., Chon, J. W. M., and Gu, M. Five-dimensional optical recording mediated by surface plasmons in gold nanorods. *Nature* 459, 410–413 (2009).
[16] Li, X. P. et al. A thermally photo reduced graphene oxides for threedimensional holographic images. *Nat. Commun.* 6, 6984 (2015).
[17] Ren, H. R. et.al. On-chip on interference angular momentum multiplexing of broadband light. *Science* 352, 805–809 (2016).
[18] Deng, R. R. et al. Temporal full-colour tuning through non-steady-state upconversion. *Nat. Nanotechnol.* 10, 237–242 (2015).
[19] Appeltant, L. et al. Information processing using a single dynamical node as complex system. *Nat. Commun.* 2, 468 (2011).
[20] Paquot, Y, Duport, F, Smerieri, A, Dambre, J, Schrauwen, B, Haelterman, M, et al. Optoelectronic reservoir computing. *Sci Rep.*, 2, 287 (2012).
[21] Duport, F, Schneider, B, Smerieri, A, Haelterman, M, and Massar, S. All-optical reservoir computing. *Opt Express.* 20(20), 22783–22795 (2012). https://doi.org/10.1364/OE.20.022783
[22] Peurifoy, J. et al. Nanophotonic particle simulation and inverse design using artificial neural networks. *Sci. Adv.* 4, eaar4206 (2018).
[23] Shen, Yichen, Harris, Nicholas C., and Skirlo, Scott. Deep learning with coherent nanophotonic circuits.
[24] Patterson, Josh and Gibson, Adam. *Deep learning: A practitioner's approach* (1st ed.). O'Reilly Media, Inc, 2017. https://www.oreilly.com/library/view/deep-learning/9781491924570/

Chapter 2

AI-integrated photonic crystal fiber-based devices

Noonepalle HariPriya, Yatagiri Telugu Murali, Shashank Kumar Dubey, Sandip Swarnakar and Santosh Kumar

2.1 INTRODUCTION

The previous decade saw a tremendous transformation in the landscape of research, business, defense, and medicine with the arrival of digital technologies, big data, and deep learning (DL) [1]. The incredible advancement of DL over the past few decades has moved the field of artificial intelligence (AI) beyond primarily theoretical computer science studies and toward the development of applications that are practical, such as medical imaging [2, 3], distributed sensing [4], speech recognition [5, 6], metrology, space applications, and aviation. Mainly, AI mimics human intelligence to find recurrent patterns based on historical data and create predictive models with the use of machine learning (ML) algorithms that identify trends and foretell future events. According to recent reports, the performance of human specialists has been compared to or even surpassed by AI-generated solutions in several applications. This significant accomplishment is the result of DL, which uses methods like artificial neural networks (ANN) and deep neural networks (DNN) to identify patterns that are hidden from human perception [7, 8]. Photonics helps the arduous data collection process, which is a crucial step in the creation of AI. Additionally, AI advances photonics by improving the capabilities of optical technology, such as optical sensing, imaging, diagnosis, and communication. The applications are of use in optical sensors in transportation like autonomous cars [9]; optical sensors for diagnosis of several diseases like Alzheimer's disease, Parkinson's disease, etc. [10–15]; optical network components for optical communication [16, 17]; and optical telescopes in astronomy [18].

The field of photonics investigates how light interacts with objects that have experienced a sharp increase in attention over the past few decades and developed into a thriving study area with multiple applications, as well as fundamental studies. Due to their higher performance, low power consumption, noise reduction, and tiny die size, all-optical devices will eventually replace electrical devices in the real world. High power consumption and signal jitter are the drawbacks of electronic components [19]. To fix these issues, a relatively quick optical signal is examined as a transmission signal.

DOI: 10.1201/9781003591030-2

Instead of using electrons to transfer data, optical communication uses light. The name "photonics" suggests the use of photons (light) for data transmission and processing. To run optical networks and communication technologies, all-optical parts and circuits are necessary. High bandwidth (BW) and speed are goals of modern telecommunication networks, which are particularly challenging to achieve with electronic-based technologies [20]. As a result, engineers in the fields of optics and photonics are developing all-optical components. Consequently, photonics has become a critical paradigm for enabling optics-based microprocessors on integrated circuit chips.

This emerging technology has given rise to PDs that have been implemented in many approaches, such as electro-optical phenomenon, semiconductor optical amplifier (SOA) including SOA Mach-Zehnder Interferometer (SOA-MZI) [21, 22], SOA ultrafast non-linear interferometer (SOA-UNI) [23, 24], and four wave mixing (FWM), whose response time hangs on carrier's comeback time, which restricts the performance of PDs, photonic crystals (PhCs) [25, 26], optical fibers, and plasmonic waveguides (WGs) [27–37], which have their own advantages and disadvantages in terms of switching time, speed, power consumption, size, etc. PhCs are two-dimensional photonic crystals that can be arranged with microstructures along the length of the fiber and are also known as hole-assisted fibers, microstructure fibers, and holey fibers [38]. Low bending loss, high power supply, low dispersion over a wide range of wavelengths, minimization of the thermal gradient effect, flexibility of various filling materials, and distant sensing are all advantages of photonic crystal fiber (PCF)–based sensor designs [39]. The different structures of the PCF, like hollow core, solid core, hexagonal, octagonal, rectangular, etc., are depicted in Figure 2.1. The adaptability of PCF allows for the construction and manipulation of light utilizing different lattice types, hole sizes and shapes, and single mode over all optical ranges, making them appropriate for a variety of applications.

2.2 PCF-BASED ALL-OPTICAL DEVICES

In the designing of Arithmetic and Logic Unit, measuring electric and electronic parameters, data transition, telecommunications, and all-optical digital communication systems, all use various combinational of PDs, such as logic gates (LGs) [41–44], adders, subtractors, multiplexers (MUX), demultiplexers (DEMUX), encoders, decoders, comparators, converters, signal processing devices, and sequential circuits [45]. Size, lattice constant (a), radius (r), operating wavelength (λ), photonic band gap (PBG), switching threshold (S_{Th}) measured in KW/m2, rise time (T_r), fall time (T_f) measured in pico seconds (ps), form, and contrast ratio (CR) have all been examined and compared for these devices.

Numerous articles have described the design of all-optical AND gate structures in a variety of lattice types, including hexagonal [46, 47],

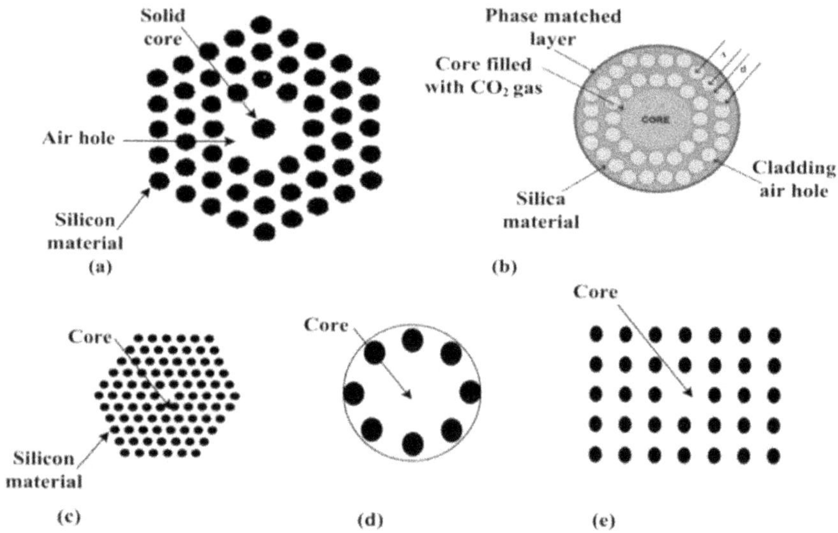

Figure 2.1 Different structures in PCF: (a) solid core, (b) hollow core, (c) hexagonal, (d) octagonal, (e) rectangular [40].

rectangular [48], elliptical [49], and triangular [50], which operate at a wavelength of 1,550 nm and make use of the finite difference time domain (FDTD) method [51, 52]. T-shaped PhC LGs performing several operations in single layout with non-linear (NL) material [53] and linear material [54–59] have been designed by shifting the phase angles at the inputs. Full adders (FA) are created by PhCs employing T-shaped WGs, Rsoft photonics simulation software, and the BANDSOLVE and FULLWAVE toolboxes for band diagram and wave propagation, respectively [60–62]. To get over the constraints of earlier papers [63, 64] and the Galois field adder [65], we used four Ex-OR gates; these are likewise constructed with Y-shaped WGs without the need of NL material or SOA. The all-optical half subtractor (AOHS) based on PhC is implemented, which includes two optical switches that help with dropping operation, one power splitter, and one power combiner [66]. The novel passive full subtractor (FS) using PhCs is realized by using linear material [67] and NL material [68] made of doped glass (DG) [69] by using the concept of beam interference sequence. The structures [70, 71] perform both operations of MUX/DEMUX made of different WGs and Ring Resonators. A comparator is employed using linear the material of T-shaped WG [72] and NL material with Y -shape in [73] and RRs in [74, 75]. The parameters of different types of PDs like LGs, combinational circuits, etc., without using AI have been compared and summarized in Table 2.1. The AND gate designed in [46, 52] had three inputs, while all the other gates had two inputs. The circuits in [52, 53, 59] were designed with small layouts, which have a low size compared to other designs that can be used in high-speed integration circuits. The circuits have been designed with many NL materials, such as gallium arsenide, DG, Kerr effect, and chalcogenide, that shift the wavelength and frequency of the light. The PBG is the main parameter of the PhC, which stops the entry of light in some regions, which is summarized from several articles in Table 2.1. The wavelength followed by several structures designed using PhCs is at a range of 1,550 nm. The main parameters that are used to design the circuit are r and a, which will be at a range of 120 nm and 600 nm, which are followed by many articles.

2.3 PCF-BASED SENSORS

PhC biosensors offer a wide range of uses in biomedical research, health care, pharmaceuticals, and other applications. The optimum performance in biosensor applications is displayed by a PCF. A two-dimensional square lattice PhC with a line defect is employed as the glucose sensor in [76] to measure different urine glucose concentrations. The photon passes through PhC at a glucose concentration in order to react with urine at one end and the other. Ethylenimine acids, polyacrylic, and Single-Stranded DNA are just a few of the solutions used to detect DNA. At a level of 1 µm, the SS-DNA's accuracy value is 0.27 nm. Several devices and sensors that were designed using PhCs are shown in Figure 2.2. In a similar way [80], gold electrodes at a concentration of 1 µm were immersed in a pH 7.4 solution to carry out the SS-DNA. Chloroform is considered for analysis due to its high RI dependency on temperature and broad transmission window—i.e., visible to the mid-infrared region that improves the temperature detection range [81]. A PCF sensor with a D shape operates in the visible to near-infrared band to provide minimal loss and extremely high sensitivity; this structure is therefore simply implementable for fabrication and is suitable for a variety of sensing purposes [78]. Due to the properties of PCF, these are exploited in biomedical [40, 82], physical [83–87], biochemical [77, 88–90], and agriculture [79] applications.

2.4 SYNERGY BETWEEN PHOTONICS AND AI: DATA COLLECTION, OPTICAL TECHNOLOGY, AND ALGORITHMS

Researchers from all over the world are very interested in finding out how to use AI for photonics. The terms "artificial intelligence," "deep learning," or "machine learning," and "photonics" were used to search the number of publications from the Web of Science that cover these topics and others from 1996 to 2021. There was enormous growth in publications during 2005, from 5,000 to approximately 50,000, and it is anticipated that this upward trend will continue for years.

Table 2.1 Comparison of several parameters of all-optical photonic devices (PDs)

Device	Shape	Material used	No. of inputs	Size (um2)	a (nm)	R (nm)	λ (nm)	PBG	CR (dB)	RT (ps)	TE (%)	IL (dB)	BR (Tbps)	Other parameters	Ref
AND	YWG	Linear	3	158.76	600	120	***	1,250–2,142	10.96	***	***	***	***	***	31
AND	YWG	Linear	2	***	420	168	1,550	1,042–1,720	6.13	***	***	***	0.664	***	32
AND	YWG	Linear	2	104	540	99.9	1,550		***	***	***	***	***	***	33
AND	YWG	Linear	2	396	750	215	1,550	1,388–1,666	***	0.56	***	***	***	***	34
AND	YWG	Linear	2	540	600	210	***	***	***	***	***	***	***	***	35
AND	YWG	NL (Gallium Arsenide (GaAs)	2		384	***	1,550	***	***	0.4	***	***	***	***	36
AND	TWG	Linear	3	45.36	600	120	1,550	856–1,253	24.53	0.026	19.6	1.52	38.4	***	37
All-optical gate	TWG	NL (GaAs)	2	74.30	575	155.25	1,550	***	***	***	***	***	***	***	38
XOR, NOT, OR	TWG	Linear	2	315	600	120	1,550	1,440–2,120	55.31	0.1456	***	***	6.868	***	39
XOR, NOT, OR, AND	TWG	Linear	2	315	***	96	1,550	1,440–2,120	54.91(XOR, NOT) 6.02 (AND)	0.15	***	***	***	***	40
NOT/XOR, OR/XOR	TWG	Linear	2	72.8	***	***	***	***	62.38	0.235	***	***	***	***	41
AND, OR, XOR	TWG	Linear	2	95.76	600	120	1,550	***	33.05 (AND) 10.5 (OR) 8.29 (XOR)	***	***	***	***	***	42

NAND, NOR, XNOR	***	Linear	2	38.88	***	***	***	***	17.59 (NAND), 14.3 (NOR) 10.52 (XNOR)	***	***	***	***	***	43
XOR, XNOR	TWG	Linear	3	38.88	600	120	1,550	850–1236	13.3 (XOR) 11.8 (XNOR)	12	***	***	83.33	***	44
HA	Hexagonal resonator	***	***	150.48	650	104	1,530	1,250–1,910	***	***	***	***	***	***	45
FA	TWG	NL (Chalcogenide)	***	71	461	135	1,550	1,380–1,780	13.22 (XOR), 13.84 (AND)	0.39	***	***	***	***	46
FA	TWG	***	***	72	409	109	1,550	1,136–1,565	11	2	***	***	0.5	***	47
FA	YWG	Linear	***	345.6	600	120	1,550	***	***	3.5	***	***	***	***	48
FA	YWG	***	***	362.52	640	115.2	1,550	1,280–2,133	***	1.5	***	***	***	***	49
HS	QS resonator	NL (Kerr Effect)	***	602	630	226	1,547	1,432–1,969	***	1.4	***	***	***	***	50
FS	YWG	Linear	***	465	600	120	1,550	1,428–2,097	3.97	0.16	***	***	6.25	***	51
FS	Square resonator	NL	***	368.28	580	116	1,550	1310–1,870	***	***	***	***	***	Tr = 2.8 Tf = 0.2	52
FS	Square resonator	NL	***	1592	600	120	1,550	***	9.8	***	***	***	***	Tr = 0.9 Tf = 0.7	53
MUX/ DEMUX	Sharp bend WG	Linear	2*1	71.28	***	***	***	***	***	***	***	***	***	***	54

(Continued)

Table 2.1 (Continued)

Device	*Shape*	*Material used*	*No. of inputs*	*Size (um2)*	*a (nm)*	*R (nm)*	*λ (nm)*	*PBG*	*CR (dB)*	*RT (ps)*	*TE (%)*	*IL (dB)*	*BR (Tbps)*	*Other parameters*	*Ref*
MUX/ DEMUX	TWG Coupling	Linear	3*1	***	601	120	1,550	1,250–1,810	***	***	***	***	***	***	55
Comparator	TWG	Linear	1-bit	132.25	***	***	1,500	790–1,150	***	***	***	***	***	***	56
Comparator	TWG	NL (GaAs)	1-bit	***	***	***	1,550	1,230–2,000	***	0.07	***	***	3.33	***	57
Comparator	Square Resonator	NL (DGI)	1-bit	1,705	600	126	1,550	1,500–2,200	***	6	***	***	***	***	58
Comparator	X-shaped Resonator	NL (Kerr Effect)	1-bit	725.4	600	120	1,550	1,300–2,222	***	***	***	***	***	Tr = 1.5 Tf = 2	59

*** Not available

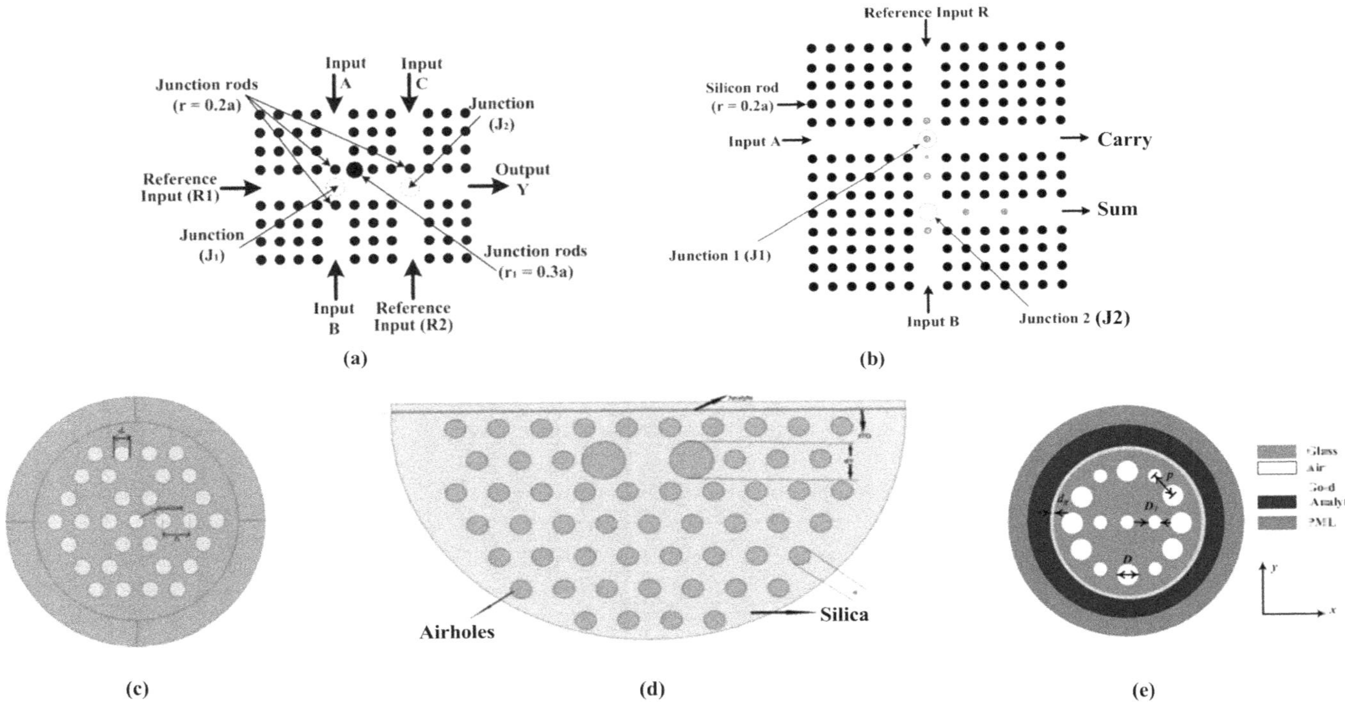

Figure 2.2 PDs and sensors designed using PhCs. (a) Diagrammatic representation of the XOR/XNOR layout [59]. (b) Schematic layout of proposed half adder circuit [52]. (c) The simulated cross section of LMA-10 [77]. (d) Schematic of the proposed structure [78]. (e) Two-dimensional view of the proposed PCF-based Surface Plasmon Resonance sensor [79].

DL- and AI-related photonics articles have dramatically expanded recently, such as artificial, convolutional, regenerative, generative NNs, regression [91, 92], and clustering [93–95]. A number of studies [96–98] examine photonics applications of AI approaches. Others [96, 97] focus on using AI techniques for general PDs, while one [98] only uses AI techniques for integrated quantum photonics. The flowchart depicted in Figure 2.3 shows several techniques that have been linked with AI and devices that can be designed using PhCs, which, combined, are ruling the present world. Some of the structures are shown in Figures 2.3(c) and (d).

2.5 AI-ENHANCED PHOTONICS: APPLICATIONS, CHALLENGES, AND OPPORTUNITIES

The usage of AI in the device modeling of PDs can be majorly classified into two types: forward modeling and inverse modeling.

2.5.1 Forward modeling

In forward modeling, the device geometrical parameters make up the input parameter space, while the performance vectors make up the output parameter space. AI techniques frequently use discriminative neural networks (NN). Through multilayered feedforward networks, or convolutional NNs (CNN), discriminative NNs simulate the precise mapping between the input and output variables. The spectral response of these has been effectively modeled in advance using electric field profiles of photonics systems [101, 102], plasmonic scatters [103], dielectric metagratings [104], topological insulators [105], effective RI of WGs [106], dielectric metasurfaces [107–109], and beam combiners [110].

Photonics design tasks are traditionally based on physical principles and intuitions, which limit the exploration of the input parameter space. Maxwell equations are the governing equations for photonics modeling and simulation, but they are computationally expensive and laborious to solve numerically. Soft computing algorithms based on ML can potentially discover novel and optimal designs that are beyond physical intuitions, and accelerate the execution speed by orders of magnitude. Soft computing algorithms can also incorporate uncertainties in material parameters, fabrication, and manufacturing, and reduce fabrication tape-outs and cycles, and increase photonics manufacturing yield. Optical mode solving is an area where soft computing algorithms can be applied to improve the design and modeling of electromagnetic problems.

The design of photonics integrated circuits requires optical mode solving, which is the process of finding the solutions to the time-independent Maxwell's equations for optical waveguides. These solutions provide information about the mode profiles, field confinements, and effective and group

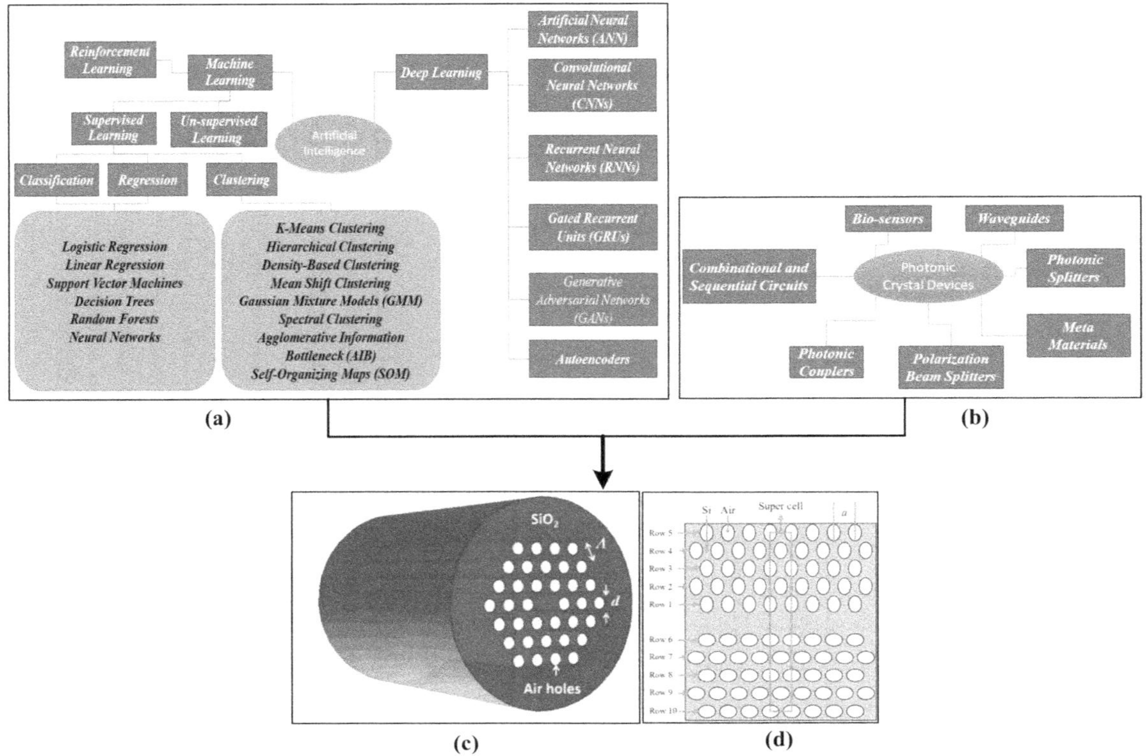

Figure 2.3 (a) Domains related to AI. (b) Devices that can be designed using PhCs. (c) Schematic of the index-guiding, microstructured optical fiber [99]. (d) Proposed asymmetric oval shaped, air-hole photonic crystal waveguide (AOPCW) structure with super cell [100].

RIs, which are important for designing more complex components. However, solving Maxwell's equations numerically is a difficult and time-consuming task, especially for two-dimensional geometries, such as channel and strip waveguides. Traditional numerical methods, such as finite difference and finite elements, are based on matrix-diagonalization techniques, which consume a lot of computational resources. Moreover, these methods are inefficient, as they do not take advantage of the patterns that can be learned from exploring the parameter space of the waveguide geometry. The parameter space is usually well-defined by the choice of available materials and fabrication capabilities, and researchers often repeat the same calculations for different values of the parameters. These repeated calculations can be avoided by learning the patterns from the previous calculations and using them as an effective representation for future calculations without the need of numerical methods.

The following sections highlight recent efforts on optical mode solutions using soft computing approaches. We demonstrate modal classification, effective N calculations, and optical mode profile predictions using DL algorithms. A DL model for waveguide model classification is provided in [111] and effective RIs in [112]. Consideration is given to silicon (Si) nitride channel WGs with silica cladding. The modal classification is based on a densely connected feedforward architecture with four hidden layers, where w and h are represented by the horizontal and vertical axes, respectively. The exact single-mode curve produced using the standard FDTD is represented by the solid black line B(w,h). This curve divides the input parameter space into single and multimode zones.

2.5.2 Inverse modeling

In inverse modeling, on the other hand, the inputs are the performance vectors, and the outputs are the corresponding device geometry, as the name implies. AI approaches in inverse modeling are further classified into three types.

The first category employs gradient descent algorithms in conjunction with a forward model. The gradients of the forward model are evaluated in this case using either automatic differentiation [113, 114] or adjoint methods [115, 116]. Automatic differentiation employs derivative chain rules, which are equivalent to back-propagation algorithms technically. Starting with an informed guess of the device geometry, the error between the guess geometry response and the required response is calculated. This error is propagated backward, resulting in a modified geometry. The practice is repeated until the gap between the guess and the intended outcome is acceptable. The second type of inverse method employs traditional optimization techniques, such as genetic and particle swarm algorithms, as well as forward models. The third type of inverse algorithm employs generative NN to forecast device geometry while meeting performance requirements [117–119].

The three most frequent NNs are CNN, recurrent neural network (RNN) and ANN. ANN is typically composed of three layers: input, hidden, and output, all of which are connected in the forward direction. Each node in the input and hidden layers includes weight and bias settings for the node connecting to the next layer. ANN can learn sophisticated NL functions that can map any input to the desired output. Based on various measured data, [120] projected hourly global, diffuse, and direct sun irradiance using ANN. However, the size of the input data (the number of nodes in the input layer) and the number of nodes in the hidden layer must be relatively small for ANN to be useful because as these numbers rise, the number of trainable parameters (weights and bias) also rises sharply. Additionally, ANN cannot capture sequential information in the input data, such as time series data, and loses the spatial/temporal aspects of the input data, such as photos and video. In the DL community, particularly in computer vision, CNN is widely used. The trainable parameters for filters, also known as kernels, are different from ANNs. Instead of fully connecting all the nodes in each layer, the filters are utilized to convolutionally extract the pertinent information from the input. For instance, [121] employed CNN to identify COVID-19 symptoms by extracting high-level features from computed tomography data. In natural language processing with sequential input data, RNN is frequently utilized. It shares parameters over many time steps and has a recurrent link on the hidden layers. As a result, there are fewer parameters to train with and the computational cost is reduced. An RNN is used in [122] to determine the field continuity between adjacent pixels for the purpose of calculating optical mode profiles.

Numerous frameworks and approaches for ML are based on the three fundamental NNs; several applications, including generative adversarial networks (GAN) [123, 124], ResNet [125], Transformer [126], and LSTM [127, 128]; and solving a variety of challenging issues. DNNs outperform established methods in discovering new photonics structures from massive data to achieve optimal optical performance. In this section, we review deep generative models (DGMs) for the inverse design of photonics structures, providing some insights into how these deep generative models can solve the inverse design problems. DGMs are methods that integrate generative models with DNNs and have achieved remarkable success in a short period of time. The DGMs use DNNs to develop a function capable of approximating the model distribution to the genuine data distribution of the training set, resulting in fresh data points with some variances. Popular DGMs include GANs [129], variational autoencoders (VAEs) [130], and auto-regressive models [131], with the former two being the most widely employed. GANs have been used to build inverse metasurface nanostructures [132] and dielectric metasurfaces [133]. VAEs have been used to build the inverse of double-layered chiral metamaterials [134]. Figure 2.4 shows the clear description of the forward model and the inverse model where the forward model is like choosing a device structure, such as the empirical structure,

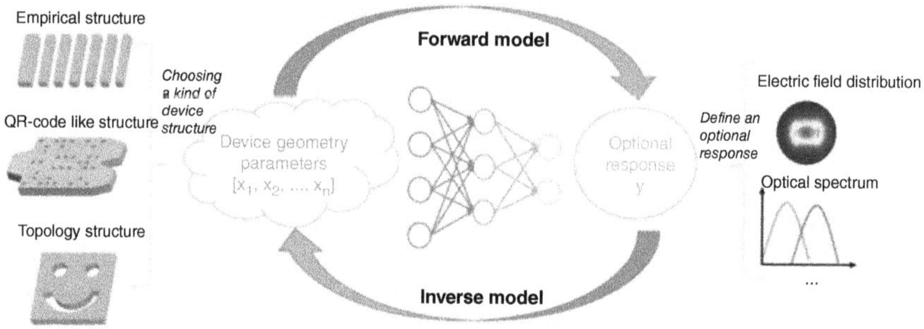

Figure 2.4 Schematic of DNN-assisted silicon photonic device design approach [135].

Q-R code–like structure, or topological structure, and then analyzing the parameters such as electric field distribution, Optical spectrum, chromatic dispersion, etc., through the neural networks whereas the inverse modeling will define an optical response initially and then the device is structured using the NNS.

2.6 PDs USING AI

Many electro-optical and PDs, including metasurface, plasmonic MIM WGs [136–140], and lithium-niobate-based MZI [141], were used to create LGs as photonics and semiconductor fabrication technology advanced. These tools can also be used to execute different LGs. In conclusion, as PhCs have characteristics like high operation speed, small size, low cost, and high efficiency, it is also important to create NN using a PhC approach to improve operation speed and efficiency.

2.7 LGs USING INVERSE MODELING

All major LGs, including OR, AND, NOT, NOR, XOR, XNOR, and NAND, are adequately created by employing the linear interference approach, including the creation of topological filter and all-optical LGs based on two-dimensional PhCs with durable edge states [142]. The incident plane wave is first spatially encoded by a particular logic operation (NOT, AND, OR) at the input layer and then further decoded through the hidden layers—namely, a compound Huygens' metasurface—using a simple yet effective design strategy that can be used with plane waves as the incident signal. The diffractive NN is trained on a server with a GeForce 249 10 GTX Titan X GPU and operating system using Python version 3.5.0 and the TensorFlow framework version 1.10.0 [143]. Both supervised and unsupervised learning techniques are used to study the LGs of optical NN (ONN). The proposed structures perform OR and XOR operations for the optical neurons make use of both interconnection configuration and self-connection configuration [144]. ONNs can implement neuron functionality and interconnection using optical and PDs and the physics of light propagation. To conduct optical logic operations, an on-chip diffractive optical neural network (DONN) is used. The encoded light at the input layer is decoded through the hidden layers (One Dimensional - metasurfaces) in this configuration. Each meta-atom functions as an independent neuron in the NN, interconnecting with other meta-atoms in the neighboring layers via light diffraction [145].

2.8 LGs USING FORWARD MODELING

There are two main approaches to all optical LGs: One is based on the linear optical effects, and the other is based on the NL optical effects, especially the third-order NL susceptibility [146]. Using a spin-encoded technique, we were able to create XNOR LGs using an optical slot nanoantenna with a footprint of 220 nm by 60 nm and OR, AND, NOT, NAND, and NOR LGs using an L-shaped optical slot nanoantenna with a footprint of 300 nm by 300 nm. A single element nanoantenna integrates the surface plasmon polaritons launch and logic operation via mode coupling rather than path interference, greatly reducing the device's size. In subwavelength, all-optical logic computing and nanophotonic information processing, the experimental results demonstrate the promise of the single optical slot nanoantenna [147]. Two-dimensional NL photonic crystal slabs are used to demonstrate the theoretical design of ultracompact all-optical AND, NAND, OR, and NOR gates. NL material is a compound Ag-polymer film with a low refractive index and significant third-order nonlinearity. PhC cavities with a comparatively high-quality factor of roughly 2,000 are constructed on this polymer slab [148]. Multimode interference WGs made of PhC are the foundation for all-optical XOR, XNOR, NAND, and OR functions that have been designed [149]. Figure 2.5 shows some structures of the LGs designed using AI techniques in which every structure follows forward modeling. In a similar way, the comparisons of several parameters of different LGs designed using integrated AI are summarized in Table 2.2, which has the highest bit rate in [145], where the structure in it performs operations of OR, XOR, XNOR, NAND.

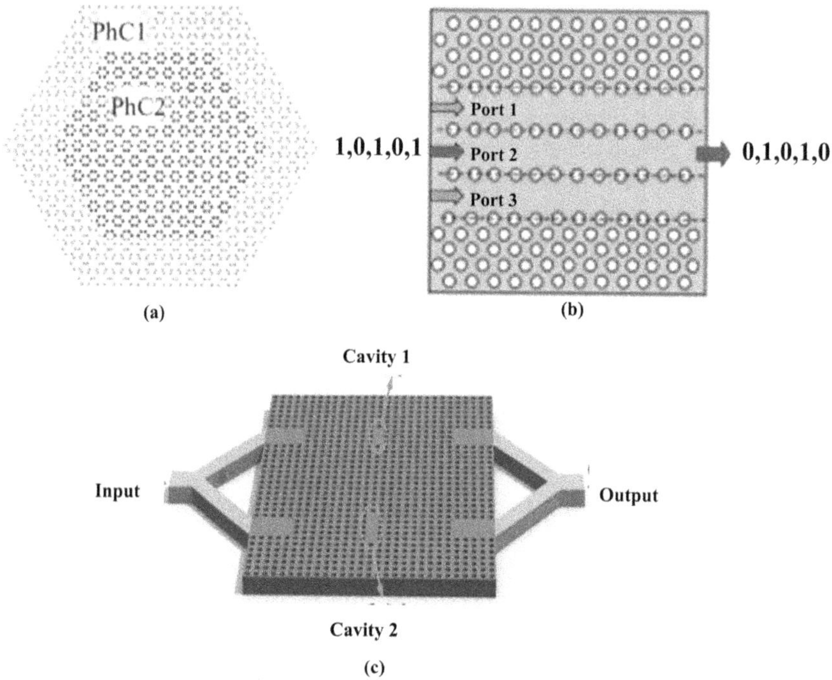

Figure 2.5 LGs designed based on the AI-integrated PhCs. (a) Schematic of topological cavities (topological PhC1 outside, trivial PhC2 inside) [142]. (b) Realization of a functional all-optical NOT logic gate [146]. (c) Schematic structure of two-dimensional PhC slab for all-optical LGs [148] and the smallest design in [149].

2.9 PARAMETERS ANALYZATION OF PDs USING FORWARD MODELING

The analysis of PCFs has been transformed by forward modeling techniques [99]. Notably, chromatic dispersion prediction has been a crucial component of PCF design. Traditional numerical techniques, such as finite element analysis, require a significant amount of computer work. ANNs have become effective workarounds to this problem. Based on PCF parameters, they use previous data to forecast chromatic dispersion [150]. Additionally, beyond the dataset, NNs demonstrate promise in forecasting topological transitions in PhCs [151]. These networks exhibit improved generalization ability by using physical principles derived from Maxwell's equations. The performance of the PhC MZI is also improved through NNs optimization. Refinement of MZI structural parameters using back-propagation NNs results in a striking rise in normalized transmission peaks [152]. Radial basis function-based artificial neural networks (RBF-ANNs) have been

Table 2.2 Comparison and summarization of several parameters obtained for AI- integrated LGs

Structure	*Shape*	*Realized method*	*size*	*A*	*r*	*Wave (um)*	*Band width*	*CR*	*Bit rate (Gbps)*	*Other parameters*	*Ref*
OR, AND, NOT, NOR, XOR, XNOR, and NAND	Two-dimensional PhC	Topological PhC's cavity and WGs	2,250	1	0.12	2.5793	***	54.4 (OR, XOR, NOT), 32.6 (XNOR) 31.1 (NAND) 9.54 (NOR, AND)	***	***	124
NOT, AND, OR	Two-dimensional PhC	DNN	1,260	***	***	***	***	9.6	***	17 GHz	125
OR, XOR	Two-dimensional PhC	Supervised and Unsupervised Techniques	***	***	***	***	***	***	3	Length (L) of the optical fiber is 0.06811 m, RI of silica 1.46 delay time (T_D) = 0.33 ns	126
AND, OR, NOT	Two-dimensional PhC	On-chip DONN	18,500	500	***	1.55	60	>10.82	20	***	127

(Continued)

Table 2.2 (Continued)

Structure	*Shape*	*Realized method*	*size*	*A*	*r*	*Wave (um)*	*Band width*	*CR*	*Bit rate (Gbps)*	*Other parameters*	*Ref*
AND, NOT	Coupled NL PhCs	***	***	***	***	***	***	20	***	***	128
AND, NOT, NAND, XOR	L-shaped optical slot nano resonator	Surface plasmon polaritons	90,000	***	***	800	230 nm	>13.27	***	***	129
AND, NAND, OR, NOR	NL PhC slabs using line defect cavity	***	***	***	***	***	***	***	***	***	130
XOR, XNOR, NAND, OR	PhC multimode interference WGs	***	46.23	***	***	***	***	28.6 (XOR, XNOR), 25 (NAND), 26.6 (OR)	***	Tolerance −10% to 30 %	131

***Not available

shown to be superior to traditional techniques in comparative examinations of NN topologies [153]. RBF-ANNs outperform CNNs, whether they use Gaussian or another type of basis kernel function. The precise study of PCFs is further performed using the radial basis function method [154]. For increased accuracy, this strategy uses calculated data after training and calculation processes.

The modeling of non-Markovian dynamics in PhCs receives a major boost. RNNs employ gated recurrent unit (GRU) layers to speed up prediction and cut down on training time [155].

2.10 PARAMETERS ANALYZATION OF PDs USING INVERSE MODELING

In the design and study of photonic WGs, inverse modeling has been applied in revolutionary ways. For instance, DL algorithms provide effective substitutes for precise guided mode predictions in photonic WGs. Utilizing input characteristics like breadth (b), height (h), and wavelength, these techniques have been used to determine the effective modal indices of Transverse Electric and Transverse Magnetic modes [156]. DNNs and iterative optimization methods like Genetic Algorithm (GA), Particle Swarm Optimization (PSO), and Direct Binary Search (DBS) have demonstrated effectiveness in the field of inverse design for Si photonics. Over a broad wavelength range, subwavelength structures created using the GA algorithm have shown transmission efficiencies of both TE and TM modes over 85%. To reduce the need for voluminous training data, semi-supervised generative NN such as Conditional Variational Autoencoders (CVAE) and Conditional Generative Adversarial Networks (CGAN) have been developed [135]. The computation of power coupling efficiency in photonic coupler devices has been enhanced by the use of the multilayer perceptron as an alternative to finite element techniques (FEM) [157]. DBS and GA are effective methods for designing photonic elements [158]. Additionally, generative models like CNNs and GANs make high-dimensional data representation possible. By sampling latent spaces, this method facilitates the creation of various models [159]. DL is essential for mapping the design space of photonic topological states, and DNNs are advantageous for dispersion-oriented inverse design when it comes to predicting structural parameters [160]. Additionally, NN based on genetic algorithms are helpful for designing extremely efficient PDs [161]. Genetic algorithms have been effectively used to design two-dimensional PhCs [162]. ML approaches are used to improve the design of gas sensors using PhCs [163]. The final step is the development of asymmetric, oval-shaped-hole PhC WGs using AI optimizers [100, 164]. Different parameters considered for design of the PDs using AI techniques are like chromatic dispersion, RI (RI), diameter of air holes etc. Some of the structures designed using these parameters are illustrated in Figure 2.6.

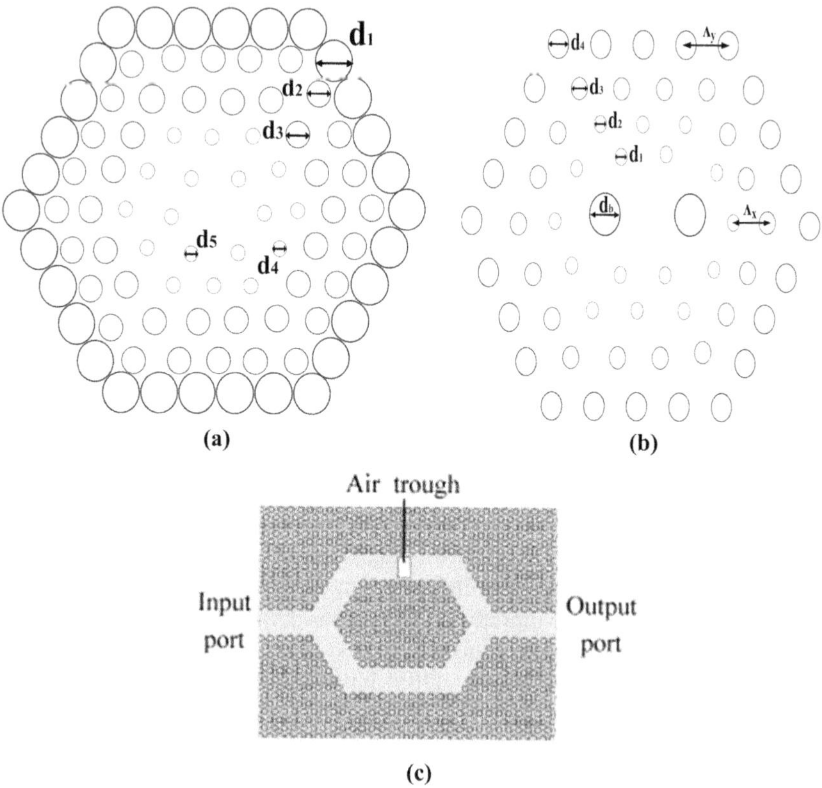

Figure 2.6 Parameters analyzation PDs designed using AI-integrated PCF. (a) Schematic diagram of ultra-flattened, zero-dispersion PCF [152]. (b) Cross section of the suggested highly birefringence PCF. The diameters of the holes in each ring are d_1, d_2, d_3, and d_4. However, d_b is the diameter of the two bigger holes in the first ring. Λ_x and Λ_y are the hole pitch in the x and y direction, respectively [153]. (c) MZI-type structure based on PC waveguide single-arm etching through structure [151].

The parameters calculated and used in the design of these PDs are explained in Table 2.3, such as chromatic dispersion, zero ultra-flat dispersion, optical mode properties, wavelength, and RI. The article [150] has the highest efficiency and less error, and the methods used in the article are physics-adapted NNs. In order to design the PCF structure in inverse modeling, several structural parameters need to be considered, such as the height, width, pitch of the air hole, and radius, to obtain less error by following several methods, like multilayer perception [99], back propagation [151], RBF ANN [135], ONN, D<<, GANN, etc.,

Table 2.3 Comparison and summarization of several parameters analyzed using AI techniques

Parameters	*Methods*	*Input Parameters*	*Output*	*Efficiency*	*Error*	*Ref*
Chromatic Dispersion (CD)	Multilayer perceptron (MLP)	Λ = [1.2μm, 1.8μm], hole diameter (*d*) = 0.4μm, 2.4μm, pitch hole (Λ) = 2μm, 3μm	CD = 238, 140 ps nm-1 km-1	***	8×10^{-5}	132
Topological invariant of photonic bands	Physics-adapted (NN)	Operator parameter space (encoded with Maxwell's equations in k-space)	Topological invariant of photonic bands	Accuracy rate 99.72% for the training dataset and 97.32% for the test dataset	< 0.5%	133
MZI structure based on triangular lattice PC WG	Back-propagation (BP) NN	Silica RI (*nSi* = 3.5), a = 470nm, size = 280.8 μm^2 Filling rate = 0.74, λ = 1,550 nm, concentration of salmonella solution 10^2–10^8 cuf/mL,	Transmission spectrum peak wavelength shift = 1,552.5 nm to 1,551.3 nm (single-arm notching structure), 1,552.4 to 1,560.1 nm (double-arm notching structure), sensitivity = 4,583nm/RIU (double-arm notch), Transmission peak of MZI increased from 85.3% to 97.1%	***	10^{-4}	134

(*Continued*)

Table 2.3 (Continued)

Parameters	*Methods*	*Input Parameters*	*Output*	*Efficiency*	*Error*	*Ref*
Polarization rotator based on slanted rib waveguide	RBF ANN	h = 1.2μm, b = 2.3μm, second substrate height (t) = 0.1μm, substrate RI (n_s) = 3.27μm, cladding RI (n_c) = 1.0μm	Maximum conversion power (Pmax), conversion length (LC), ratio of Pmax/LC (R), field profiles of quasi-TE and TM modes, TE and TM mode power variations	***	Trained data AT λ = 1.55μm, 0.145% (RBF ANN) and = ±0.235% (CNN), At λ = (1.0μm –2.0μm) ±0.17 (RBF ANN), ±0.28% (CNN) untrained data- At λ = 1.55μm, ±0.779% (CNN), ±0.773% (RBF ANN) At λ = (1.0μm to 2.0μm) ±1.0% (RBF ANN and CNN)	135
Zero and ultra-flat dispersion PCF	RBF ANN	d = d_3[0.7, 0.8384, 0.9μm], d_2[0.5μm, 0.6157μm, 0.7μm], d_1[0.4μm, 0.5381μm 0.6μm]), λ = (1.0μm – 2.0μm)	Effective index (n_{eff}), Chromatic dispersion	***	7.6 ×10%–6% (CNN), 2.0 × 10%–7% (RBF ANN)	

Mode properties of PCF	RBF ANN	Training data with full vectorial finite difference method (FVFDM)	Effective index (n_{eff}) of fundamental mode, effective index of fundamental space-filling mode (n_{FSM})	***	n_{eff} [At d/Λ = 0.6 for 16 neurons = 1×10^{-6}%, 32 neurons = 0.3 $\times 10^{-6}$%, 64 neurons = 0.17 $\times 10^{-6}$%] [At d/Λ = 0.63 for 16 neurons =3%, 32 neurons = 0.04%, 64 neurons = 0.005%]	136
Optical mode solving	DNN	W= (0.15μm<w<1.1μm), h = (0.1μm<h<0.5μm), Core RI (1.6<n<4), Prediction domain with square (L = 1.5 μm),	Mode images	***	3.7×10^{-5}	138
Fundamental guided mode indices in photonic WGs	ONN	Si strip w = 400nm − 1400nm, h = 200nm − 400nm	Effective modal indices n_{TE} and n_{TM}	***	At λ = 1350 nm 4.5×10^{-4} (TE) 5×10^{-3} (TM) At λ = 1650 nm 3.62×10^{-4} (TE) 4.6×10^{-3} (TM) (Mean square error ~10^{-6})	139

(Continued)

Table 2.3 (Continued)

Parameters	*Methods*	*Input Parameters*	*Output*	*Efficiency*	*Error*	*Ref*
Micro-to-nano coupler waveguide	MLP, NN, adapted evolutionary strategy algorithm	Input (w − 2μm) positioned left, output (w − 0.2μm) positioned right, distance(ds) = 4.2μm, material columns (d = 0.2μm or 0.3μm), λ = 1.55μm, electric field amplitude = 1 (V/m)	Coupler 1 efficiency = 73.94%, Coupler 2 efficiency = 95% (ESCH)	***	MLP error about 1.73%	141
Optical Passive elements simulation using ANN	ANN	Optical WG [Si WG width (w_g = 200 – 1500 nm), Si thickness (th_g = 200 – 500 nm), slab th (t_s = 0 – 100 nm), λ = 1400 – 1600 nm, radial WG [Si layer th = 220 nm, buried oxide layer th = 2 μm, r = 0 to 5 μm, directional coupler [L_c= 0 to 400 μm, coupler gap (g_c = 50 nm to 300 nm)], multimode interference coupler λ = 1.55 μm, h = 220 nm, w_c = 3 μm to 6 μm]	Fundamental modes, power splitting efficiency (90%), transmission (87%)	***	Optical WG [average training error (ATE) = 0.67%, average validation error (AVE) = 0.6%], Radial waveguide [ATE = 0.32%, AVE = 0.76%], directional coupler [ATE = 0.37%, AVE = 0.66%,], multimode interference coupler [ATE = 0.53%, AVE = 1.62%]	142

Design space of photonic topological space	Forward and Inverse model estimation using trained (DNN) and (CNN) architecture	DNN (4-8-16-32-55 fully connected layer-based NN), CNN (one convolutional layer connected to four fully (320-64-32-4) connected layers)	Approximate mapping function, band gap structures	***	DNN (test error order of 10^{-3}), CNN (test error of 0.014/sample)	144
PCF for wavelength conversion	DNN	Λ = 0.5 μm, Λ = 2 μm to 5 μm, d/Λ = 0.25 to 0.45, pumped λ = [852 nm, 976 nm, 1064 nm, and 1310 nm]	Converted λ predicted by DNNs are averaged, which are 1.406 nm, 2.38 nm, 2.494 nm, and 1.41 nm	Prediction accuracy above 95%	MSE = 1.92 nm from design target	145
TE mode converter, and a broadband power splitter	GANN	220 nm device Si, 2–3μm buried oxide (BOX) layer, waveguide (w = 500 nm, λ = 1,550 nm), space feature size smaller than 0.2 μm, splitting ratio 2:3	transmission > 90%, growth of performance about 80%	***	***	146

***Not available

2.11 AI BEYOND PHOTONICS

Many fields of science and engineering have been completely transformed by the development of AI, particularly its subset of ML. Three areas of research in AI, quantum photonics, and beyond can be distinguished: Dedicated quantum computers are used in three different ways to speed up machine learning: (1) quantum ML [165], (2) ML to solve quantum physics problems [166], and (3) ML to create quantum devices and quantum computers. In the preceding sections, we discussed the methodology for using ML techniques for the design and optimization of PDs [167].

Unitary operations, which are created by a series of physical actions on the quantum systems, are used to control quantum systems. The physical operations required for information processing and quantum computation are made possible by quantum control in quantum devices. Large sets of unitary operations must be developed to control the devices to carry out specified operations or remain in intended quantum states in order to effectively operate quantum devices and programmable quantum computers. Given that the superposition of quantum states spans a continuous space with few general driving alternatives, this is not a simple task. Deep reinforcement learning is used by Haug et al. to describe a method for preparing a continuous set of quantum states [168]. The design of the control protocol must take into account any noises from the qubit environment or the control and readout procedures in order to achieve robust control of qubits or gate operations. When creating dynamic decoupling protocols, Wise et al. used DL to extract the noise spectrum linked to a qubit surrounded by an arbitrary bath and reduce the effect of the qubit noise environment [169]. August and Ni trained an RNN to optimize the dynamical decoupling sequences to suppress mistakes in quantum memory for certain noise models [170]. Instead of relying heavily on thorough error characterization, Kim et al. employed NN to infer the amount of probability adjustment on the measurement and increase the accuracy of noisy intermediate-scale quantum (NISQ) techniques [171].

Variational quantum-based algorithms (VQAs) [172], such as the variational quantum eigen solver (VQE), have become a promising contender to attain a real-world quantum advantage over classical algorithms [173, 174] with the introduction of cloud-based quantum computing hardware. Due to the intractability of deep quantum circuits, it is too difficult to use advanced DNN with current quantum computing platforms. A proof-of-concept demonstration of variational quantum circuits to approximate the deep Q-value function of a Deep Q-Network deep reinforcement learning was presented by Chen et al. in [175]. Pure and hybrid quantum algorithms for DQN and double DQN were investigated by Lockwood and Si [176], who discovered that both could perform reinforcement learning tasks with a condensed parameter space

2.12 CONCLUSION

AI has become a heated focus in and beyond photonics research in the past few years. As we have discussed in this chapter, photonics benefits a great deal from AI for efficient soft computing and inverse design. Meanwhile, AI, in turn, benefits from photonics by carrying out AI algorithms, such as complicated DNN using photonics components that use photons rather than electrons. We introduced the applications of AI on photonics modeling, simulation using soft computing, and inverse design based on a GAN model. We believe the connection between AI and photonics will continue to flourish in a mutually advancing manner both in photonics and beyond.

REFERENCES

[1] Goda, K., Jalali, B., Lei, C., Situ, G., Westbrook, P.: AI boosts photonics and vice versa. *APL Photon.* 5, 1–3. 2020 https://doi.org/10.1063/5.0017902

[2] Ting, D.S.W., Liu, Y., Burlina, P., Xu, X., Bressler, N.M., Wong, T.Y.: AI for medical imaging goes deep. *Nat. Med.* 24(5), 539–540. 2018. https://doi.org/10.1038/s41591-018-0029-3

[3] Topol, E.J.: High-performance medicine: the convergence of human and artificial intelligence. *Nat. Med.* 25(1), 44–56. 2019. https://doi.org/10.1038/s41591-018-0300-7

[4] Westbrook, P.: Big data on the horizon from a new generation of distributed optical fiber sensors. *APL Photonics* 5(2), 020401. 2020. https://doi.org/10.1063/1.5144123

[5] Deng, L., Li, X.: Machine learning paradigms for speech recognition. *IEEE T. Audio Speech Lang. Process.* 21(5), 1060–1089. 2013.

[6] Hinton, G., Deng, L., Yu, D., Dahl, G., Mohamed, A.-R., Jaitly, N., Senior, A., Vanhoucke, V., Nguyen, P., Sainath, T., Kingsbury, B.: Deep neural networks for acoustic modeling in speech recognition: The shared views of four research groups. *IEEE Signal Proc. Mag.* 29(6), 82–97. 2012. https://doi.org/10.1109/MSP.2012.2205597

[7] LeCun, Y., Bengio, Y., Hinton, G.: Deep Learning. *Nature* 521(7553), 436–444. 2015. https://doi.org/10.1038/nature14539

[8] Jordan, M.I., Mitchell, T.M.: Machine learning: Trends, perspectives, and prospects. *Science* 349(6245), 255–260. 2015. https://doi.org/10.1126/science.aaa8415

[9] Falcone, P., Borrelli, F., Asgari, J., Tseng, H.E., Hrovat, D.: Predictive active steering control for autonomous vehicle systems. *IEEE Trans. Control Syst. Technol.* 15(3), 566–580. 2007. https://doi.org/10.1109/TCST.2007.894653

[10] Lee, G., Nho, K., Kang, B., Sohn, K.A., Kim, D.: Predicting Alzheimer's disease progression using multi-modal deep learning approach. *Sci. Rep.* 9(1), 2019. https://doi.org/10.1038/s41598-018-37769-z

[11] Carbon, M., Eidelberg, D.: Functional imaging of sequence learning in Parkison's disease. *J. Neurol. Sci.* 248(1–2), 72–77. 2006. https://doi.org/10.1016/j.jns.2006.05.005

[12] Esteva, A., Kuprel, B., Novoa, R.A., Ko, J., Swetter, S.M., Blau, H.M., Thrun, S.: Dermatologist-level classification of skin cancer with deep neural networks. *Nature* 542(7639), 115–118. 2017. https://doi.org/10.1038/nature21056

[13] Kather, J.N., Pearson, A.T., Halama, N., Jäger, D., Krause, J., Loosen, S.H., Marx, A., Boor, P., Tacke, F., Neumann, U.P., Grabsch, H.I., Yoshikawa, T., Brenner, H., Chang-Claude, J., Hoffmeister, M., Trautwein, C., Luedde, T.: Deep learning can predict microsatellite instability directly from histology in gastrointestinal cancer. *Nat. Med.* 25(7), 1054–1056. 2019 https://doi.org/10.1038/s41591-019-0462-y

[14] Lakhani, P., Sundaram, B.: Deep Learning at Chest Radiography: Automated Classification of pulmonary tuberculosis by using convolutional neural networks. *Radiology*. 284(2), 574–582. 2017. https://doi.org/10.1148/radiol.2017162326

[15] Jiang, Y., Lei, C., Yasumoto, A., Kobayashi, H., Aisaka, Y., Ito, T., Guo, B., Nitta, N., Kutsuna, N., Ozeki, Y., Nakagawa, A., Yatomi, Y., Goda, K.: Label-free detection of aggregated platelets in blood by machine learning aided opto-fluidic time stretch microscopy. *Lab Chip* 17(14), 2426–2434. 2017. https://doi.org/10.1039/c7lc00396j

[16] Rafique, D., Velasco, L.: Machine learning for network automation: Overview, architectire and applications. *J. Opt. Commun. Netw.* 10(10), D126–D143. 2018. https://doi.org/10.1364/JOCN.10.00D126

[17] Mata, J., De Miguel, I., Durán, R., Merayo, N., Singh, S.K., Jukan, A., Chamania, M., Opt. *Switching Networking* 28, 43. 2018. https://doi.org/10.1016/j.osn.2017.12.006

[18] Bertone, G., Tait, T.M.P.: A new era in the search for dark matter. *Nature* 562(7725), 51–56. 2018. https://doi.org/10.1038/s41586-018-0542-z

[19] Yablonovitch, E.: Inhibited Spontaneous Emission in Solid-State Physics and Electronics. *Phys. Rev. Lett.* 58(20), 2059–2062. 1987. https://doi.org/10.1103/physrevlett.58.2059

[20] Ye, J.Y., Mizeikis, V., Xu, Y., Matsuo, S., Misawa, H.: Fabrication and optical characteristics of silicon-based two-dimensional photonic crystal with honeycomb lattice. *Opt. Commun.* 211 (1–6), 205–213. 2020. https://doi.org/10.1016/S0030-4018(02)01922-3

[21] Swarnakar, S., Reddy, S., Harijan, R., Kumar, S.: Design and modelling of all -optical NAND gate using metal–insulator–metal (MIM) waveguides-based Mach -Zehnder interferometer s for high-speed information processing. *Opt. Quant. Electron.* 53 (9), 1 –13. 2021. https://doi.org/10.1007/s11082-021-03153-x

[22] Swarnakar, S., Reddy, S.V., Harijan, R., Kumar, S.: Design and modelling of all-optical NAND gate using metal–insulator–metal (MIM) waveguides-based Mach–Zehnder interferometers for high-speed information processing. *Opt. Quant. Electron.* 53, 493. 2021. https://doi.org/10.1007/s11082-021-03153-x

[23] Oliveira, J.M., Sousa, F.B.D., Sousa, F.M.D., Oliviera, J.E.D., Costa, M.B.C., Costa, M.B.C.: A new system for all-optical AND logic gate on Semiconductor optical amplifier based Michelson. *INSCIT* 1–6. 2018. https://doi.org/10.1109/INSCIT.2018.8546703

[24] Oliveira, J.M., Silva, H.A.B., Sousa, F.B., Oliveira, J.E., Costa, M.B.C.: Design of all - optical AND logic gate at 20 Gb/s based on SOA -MI with optimum injection current and length of TWA -SOA, 2019.

[25] Loncar, M., Doll, T., Vuckovic, J., Scherer, A.: Design and fabrication of silicon photonic crystal optical waveguides. *J. Lightwave Technol.* 18 (10), 1402 –1411. 2000. https://doi.org/10.1109/50.887192

[26] Singh, J.J., Dhawan, D., Gupta, N.: All -optical photonic crystal logic gates for optical computing: An extensive review. *Opt. Eng.* 59 (11). 2020. https://doi.org/10.1117/1.OE.59.11.110901

[27] Abdulnabi, S.H., Abbas, M.N.: All-optical logic gates based on nanoring insulator - metal -insulator plasmonic waveguides at optical communication band. *J. Nanophoton.* 13(1), 1 –20. 2019. https://doi.org/10.1117/1.JNP.13.016009

[28] Kumar, S., Singh, L., Swarnakar, S.: Design of one -bit magnitude comparator using non -linear plasmonic waveguide. *Plasmon* 12(2), 369 –375. 2017. https://doi.org/10.1007/s11468-016-0273-7

[29] Pal, A., Ahmed, M.Z.S.: An optimized design of all -optical XOR, OR, and NOT gates using plasmonic waveguide. *Opt. Quant. Electron.* 53 (84), 1–13. 2021. https://doi.org/10.1007/s11082-021-02732–2

[30] Pal, A., Ahmed, M.Z. & Swarnakar, S.: An optimized design of all-optical XOR, OR, and NOT gates using plasmonic waveguide. *Opt Quant Electron* 53, 84. 2021. https://doi.org/10.1007/s11082-021-02732-2

[31] Swarnakar, S., Guddati, A., Reddy, S.K., Harijan, R., Kumar, S.: Performance analysis of optimized plasmonic half-adder circuit using Mach-Zehnder interferometer for high-speed switching applications. *Microelectron. J.* 111, 105040. 2021. https://doi.org/10.1016/j.mejo.2021.105040

[32] Swarnakar, S., Anguluri, S.P.K., Sreevani, A., Kumar, S.: A novel structure of all-optical optimised NAND, NOR and XNOR logic gates employing a Y-shaped plasmonic waveguide for better performance and high-speed computations. *Opt. Quant. Electron.* 54, 530. 2022. https://doi.org/10.1007/s11082-022-03911-5

[33] Sreevani, A., Charles, I., Swarnakar, S., Krishna, S.V., Kumar, S.: Design and characteristic analysis of an all-optical AND, XOR, and XNOR Y-shaped MIM waveguide for high-speed information processing. *Appl Opt.* 61(5), 1212–1218. 2022. https://doi.org/10.1364/AO.451871

[34] Charles, I., Sreevani, A., Krishna, S.V., Swarnakar, S., Sharma, P.S., Kumar, S.: Enhanced all-optical Y-shaped plasmonic OR, NOR and NAND gate models, analyses, and simulation for high speed computations. *Opt Quant Electron* 54, 330. 2022. https://doi.org/10.1007/s11082-022-03699-4

[35] Swarnakar, S., Anguluri, S.P.K., Kumar, S.: Design and analysis of miniaturized all-optical binary to gray code converter using Y-shaped plasmonic waveguide for optical processors. *Photon Netw Commun* 44, 21–29. 2022. https://doi.org/10.1007/s11107-022-00976-w

[36] Nanda, R., Rath, R., Swarnakar, S., Kumar, S.: design of all-optical directional coupler using plasmonic MIM waveguide for switching applications. *Plasmon.* 17, 2153–2159. 2022. https://doi.org/10.1007/s11468-022-01695-8

[37] Swarnakar, S., Basha, S.C.A., Azmathullah, S., Prabhu, N.A., Madhu, G., Kumar, S.: Improved design of all-optical half-adder and half-subtractor circuits using MIM plasmonic waveguides for optical networks. *Opt Quant Electron* 55, 94. 2023. https://doi.org/10.1007/s11082-022-04329-9

[38] Yu, X., Yong, D., Zhang, Y., Group, P.M.: *Handbook of Photonics for Biomedical Engineering*, 2021. https://doi.org/10.1007/978-94-007-5052-4

[39] Agbemabiese, P.A., Akowuah, E.K.: Numerical analysis of photonic crystal fiber of ultra - high birefringence and high nonlinearity. *Sci. Rep.* 10(1), 21182. 2020. https://doi.org/10.1038/s41598-020-77114-x

[40] Rachana, M., Charles, I., Swarnakar, S., Krishna, S.V., Kumar, S.: Recent advances in photonic crystal fiber-based sensors for biomedical applications. *Opt. Fib. Tech.* 74, 103085, 2022. https://doi.org/10.1016/j.yofte.2022.103085

[41] Priya, N.H., Swarnakar, S., Krishna, S.V., Kumar, S.: Design and analysis of a photonic crystal-based all-optical 3-input OR gate for high-speed optical processing. *Opt. Quant. Electron.* 53, 720. 2021. https://doi.org/10.1007/s11082-021-03374-0

[42] Rachana, M., Swarnakar, S., Babu, M.R., Swetha, P.M., Rangaiah, Y.P., Krishna, S.V., Kumar, S.: Optimization of an all-optical three-input universal logic gate with an enhanced contrast ratio by exploiting a T-shaped photonic crystal waveguide. *Appl. Opt.* 61(28), 8162–8171. 2022. https://doi.org/10.1364/AO.463845

[43] Swarnakar, S., Palacharla, V., Muduli, A., Kumar, S.: Design and optimization of photonic crystal based all-optical logic gate with enhanced contrast ratio. *Opt. Quant. Electron.* 55. 623. 2023. https://doi.org/10.1007/s11082-023-04823-8

[44] Swarnakar, S., Rakesh, N.V., Kumar, K.B. Yashwanth, K.C., Saikiran, Y., Muduli, A., Kumar, S.: Design of an ultracompact 3-input majority gate using photonic crystal. *ISSS J Micro Smart Syst.* 2023. https://doi.org/10.1007/s41683-023-00117-z

[45] Swarnakar, S., Noonepalle, H., Raju, K.S.R., Ramarao, G., Ramamurthy, N., Kumar, S.: Implementation of all-optical 3-dB and 1—dB directional coupler for switching applications. *Photon. Netw. Commun.* 45(3). 107–114. 2023. https://doi.org/10.1007/s11107-023-00995-1

[46] Subhalakshmi, G., Robinson, S.: Design and analysis of optical logic gate using two dimension photonic crystal. *Int. Conf. Current Trends Conv. Tech. ICCTCT.* 1–5. 2018. https://doi.org/10.1109/ICCTCT.2018.8551012

[47] Preeti Rani, Yogita Kalraand Sinha R.: Design and operation of photonic crystal based and optical logic gate. *Workshop Rec. Adv. Photon. WRAP.* 1–2. 2014. https://doi.org/10.1109/WRAP.2013.6917637

[48] Bhadel, K. and Mehra, R.: Design and simulation of 2-D photonic crystal based all-optical and logic gate. *Int. conf. on comp. Intell. & Commun. Netw.* 973–977. 2014. https://doi.org/10.1109/CICN.2014.205

[49] Mondal, H., Chanda, S., Sen, M. and Datta, T.: All optical AND gate based on silicon photonic crystal. *Int. Conf. Microw. Photon. ICMAP.* 1–2. 2015. https://doi.org/10.1109/ICMAP.2015.7408744

[50] Lee, K.Y., Yang, Y.P., Yang, I.C., Tsai, Y.L., Liao, H.F., Lin, Y.J., Lee, W.Y. and Tsai, Y.T.: Optical logic operation for AND Gate based on planar photonic crystal circuit. *Asia Commun. Photon. Conf. Exh. ACP*, 1–6. 2011 https://doi.org/10.1117/12.905637

[51] Danaie, M. and Kaatuzian, H.: Design and simulation of an all-optical photonic crystal AND gate using nonlinear Kerr effect. *Opt. Quant. Electron.* 44(1), 27–34. 2012. https://link.springer.com/article/10.1007/s11082-011-9527-y

[52] Rachana, M., Swarnakar, S., Krishna, S.V., Kumar, S.: Design and analysis of optical three-input AND gate using photonic crystal fiber. *Applied Optics.* 60(35), 1–7. 2021. https://doi.org/10.1364/AO.443424

[53] Noshad, M., Abbasi, A., Ranjbar, R. and Kheradmand, R.: Novel all-optical logic gates based on photonic crystal structure. *IOP Pub.* 350, 1–6, 2012. https://doi.org/10.1088/1742-6596/350/1/012007
[54] Shaik, E.H., Rangaswamy, N.: Phase interference dependent single phc based logic gate structure with t-shaped waveguide as xor, not and or logic gates. *PIERS-FALL*. 210–214. 2017. https://doi.org/10.1109/PIERS-FALL.2017.8293137
[55] Shaik, E.H. and Rangaswamy, N.: Investigation on phc based t-shaped waveguide as all-optical XOR, NOT, OR and AND logic gates. *Int. Conf. Indus. Info. Sys. ICIIS*. 1–6 2017. https://doi.org/10.1109/ICIINFS.2017.8300367
[56] Bouaouina, M.S., Lebbal, M.R., Boumaza, T.B., Bouchemat, M.: All-optical NOT/OR/XOR logic gates based on photonic crystal with low response time and high CR. *First Int. Conf. Commun., Contr. Sys. Sig. Proc. CCSSP*. 50–55, 2020. https://doi.org/10.1109/CCSSP49278.2020.9151640
[57] Rao, D.G.S., Swarnakar, S., Palacharla, V.: Design of all-optical AND, OR, and XOR logic gates using photonic crystals for switching applications. *Photon. Netw. Commun.* 41, 109–118, 2021. https://doi.org/10.1007/s11107-020-00916-6
[58] Rao, D.G.S., Swarnakar, S., Kumar, S.: Performance analysis of all-optical NAND, NOR, and XNOR logic gates using photonic crystal waveguide for optical computing applications. *Opt. Eng.* 59(5), 1–12, 2020. https://doi.org/10.1117/1.OE.59.5.057101
[59] Noonepalle, H., Swarnakar, S., Kumar, S.: Modeling and anatomization of three -input photonic crystal -based XOR/XNOR gates for photonic integrated circuits. *Results in Optics*. 11, 1–21. 2023. https://doi.org/10.1016/j.rio.2023.100417
[60] Parandin, F., Malmir, M.R.: Reconfigurable all optical half adder and optical XOR and AND logic gates based on 2D photonic crystals. *Opt. Quant. Electron.* 52(56), 1–8. 2020. https://doi.org/10.1007/s11082-019-2167-3
[61] Maleki, M.J., Mir, A., Soroosh, M.: Design and analysis of a new compact all-optical full-adder based on photonic crystals. *Optik*. 227, 1–9. 2021. https://doi.org/10.1016/j.ijleo.2020.166107
[62] Rachana, M., Swarnakar, S., Priya, N.H., Krishna, S.V., Sharma, P.S., Kumar, S.: High-speed optimisation of an all-optical half adder using a T-shaped photonic crystal waveguide with an improved contrast ratio. *Pramana* 96, 174. 2022. https://doi.org/10.1007/s12043-022-02398-9
[63] Nasab, A.M.V., Mir, A., Talebzadeh, R.: Design and simulation of an all optical full-adder based on photonic crystals. *Opt. Quant. Electron.* 51(161), 1–14. 2019. https://doi.org/10.1007/s11082-019-1881-1
[64] Swarnakar, S., Kumar, S., Sharma, S.: Performance analysis of all-optical full-adder based on two-dimensional photonic crystals. *J. Comput. Electron.* 17, 1124–1134. 2018. https://doi.org/10.1007/s10825-018-1177-x
[65] Andalib, A.: A novel proposal for all-optical Galois field adder based on photonic crystals. *Photon. Netw. Commun.* 35, 392–396. 2017. https://doi.org/10.1007/s11107-017-0756-2
[66] Mirali, A., Soroosh, M., Farshidi, E.: Ultra-fast all-optical half subtractor based on photonic crystal ring resonators. *J. OptoElectron. Nanostruct.* 5(1), 1–18. 2020.

[67] Abhishek, J., Krishnan, P., Robinson, S.: A design for an ultrafast all-optical full subtractor based on two-dimensional photonic crystals. *J. Comput. Electron.* 20, 433–441. 2021. https://doi.org/10.1007/s10825-020-01637-9

[68] Beiranvand, R., Mir, A., Talebzadeh, R.: Design and simulation of a very fast and compact all-optical Full-Subtractor based on nonlinear effect in 2D photonic crystals. *Opt. Quant. Electron.* 53(351), 1–13. 2021. https://doi.org/10.1007/s11082-021-03002-x

[69] Khajeheian, N., Jamali, J., Dindarlou, M.F., Taghizadeh, M.: An all optical full subtractor based on nonlinear photonic crystals. *Optik.* 245, 1–7. 2021. https://doi.org/10.1016/j.ijleo.2021.167751

[70] Centeno, E., Guizal, B., Felbacq, D.: Multiplexing and demultiplexing with photonic crystals. *J. Opt. A: Pure. Opt.* 1(5), 1–5. 1999. https://doi.org/10.1088/1464-4258/1/5/103

[71] Selim, R., Pinto, D., Obayya.: Novel fast photonic crystal multiplexer-demultiplexer switches. *Opt. Quant. Electron.* 42, 425–433. 2011. https://doi.org/10.1007/s11082-011-9438-y

[72] Rathi, S., Swarnakar, S., Kumar, S.: Design of one-bit magnitude comparator using photonic crystals. *J. Opt. Commun.* 1–5. 2017. https://doi.org/10.1515/joc-2017-0084

[73] Parandin, F.: Ultra-compact terahertz all-optical logic comparator on GaAs photonic crystal platform. *Optics & Laser Tech.* 144, 1–6. 2021. https://doi.org/10.1016/j.optlastec.2021.107399

[74] Fakouri-Farid, V., Andalib, A.: Design and Simulation of an all-optical photonic crystal-based comparator. *Optik.* 172, 241–248. 2018. https://doi.org/10.1016/j.ijleo.2018.06.153

[75] Surender, A., Asghari, M., Mehdizadeh, F.: A novel proposal for all-optical 1-bit comparator using nonlinear PhCRRs. *Photon. Netw. Commun.* 38, 244–249. 2019. https://doi.org/10.1007/s11107-019-00853-z

[76] Sultana, J., Islam, M.S., Ahmed, K., Dinovitser, A., Ng, B.W.H., Abbott, D.: Terahertz detection of alcohol using a photonic crystal fiber sensor. *App. Opt.* 57 (10), 2426–2433, 2018. https://doi.org/10.1364/AO.57.002426

[77] Yang, X.C., Lu, Y., Wang, M.T., Yao, J.Q.: A photonic crystal fiber glucose sensor filled with silver nanowires. *Opt. Commun.* 359, 279–284. 2016. https://doi.org/10.1016/j.optcom.2015.09.102

[78] Thenmozhi, H., Rajan, S.M., Ahmed, K.: D-shaped PCF sensor based on SPR for the detection of carcinogenic agents in food and cosmetics. *Optik* 180, 264–270. 2019. https://doi.org/10.1016/j.ijleo.2018.11.098

[79] Jahan, N., Rahman, M.M., Ahsan, M., Based, M.A., Rana, M.M., Guruswamy, S., Haider, J.: Photonic crystal fiber based biosensor for pseudomonas bacteria detection: A simulation study. *IEEE Access* 99, 1–10. 2021. https://doi.org/10.1109/ACCESS.2021.3063691

[80] Dar, T., Homola, J., Rahman, B.A., Rajarajan, M.: Label-free slot-waveguide biosensor for the detection of DNA hybridization. *App. Opt.* 51 (34), 8195–8202, 2012. https://doi.org/10.1364/AO.51.008195

[81] Ayyanar, N., Raja, R.V.J., Vigneswaran, D., Lakshmi, B., Sumathi, M., Porsezian, K.: Highly efficient compact temperature sensor using liquid infiltrated asymmetric dual elliptical core photonic crystal fiber. *Opt. Mater.* 64, 574–582. 2017. https://doi.org/10.1016/j.optmat.2017.01.011

[82] Nair, R.V., Vijaya, R.: Photonic crystal sensors: An overview. *Prog. Quant. Electron.* 34 (3), 89–134. 2012. https://doi.org/10.1016/j.pquantelec.2010.01.001
[83] Upadhyay, S., Kalyani, V.L., Charan, C.: Designing and optimization of nano-ring resonator-based photonic pressure sensor. *Proc. Intl. Conf. ICT Sus. Develop.* 408, 269–278. 2016. https://doi.org/10.1007/978-981-10-0129-1_29
[84] Bounaas, F., Labbani, A.: High sensitivity temperature sensor based on photonic crystal resonant cavity. *Prog. Electromagnet. Res. Lett.* 90, 85–90, 2020. https://doi.org/10.2528/PIERL20010204
[85] Yan, D., Meng, M., Li, J., Wang, Y.: Terahertz wave refractive index sensor based on a sunflower-type photonic crystal. *Laser Phys.* 30 (6), 1–5. 2021. https://doi.org/10.1088/1555-6611/ab8af2
[86] Pathak, A.K., Viphavakit, C., Rahman, B.M.A., Singh, V.K., A highly sensitive SPR refractive index sensor based on microfluidic channel assisted with graphene-Ag composite nanowire. *IEEE Photon. J.* 13 (2), 1–8. 2021. https://doi.org/10.1109/JPHOT.2021.3069396
[87] Pathak, A.K., Singh, V.K., Ghosh, S., Rahman, B.M.A.: Investigation of a SPR based refractive index sensor using a single mode fiber with a large D shaped microfluidic channel. *OSA Continuum*, 2(11), 3008–3018. 2019. https://doi.org/10.1364/OSAC.2.003008
[88] Mironenko, A.Y., Tutov, M.V., Sergeev, A.A., Bratskaya, S.Y.: On/off rhodamine based fluorescent probe for detection of Au and Pd in aqueous solutions. *Sens Actua. B: Chem.* 246, 389–394. 2017. https://doi.org/10.1016/j.snb.2017.02.092
[89] Liu, H., Wang, H., Zhang, W., Chen, C., Wang, Q., Ding, Y., Tang, S.: High sensitive methane sensor with temperature compensation based on selectively liquid in filtrated photonic crystal fibers. *Photon. Sens.* 9 (3), 213–222. 2019. https://doi.org/10.1016/j.snb.2017.02.092
[90] Habli, O., Bouazzi, Y., Kanzari, M., Gas sensing using one-dimensional photonic crystal nanoresonators. *Prog. Electromag. Res. C.* 92, 251–263. 2019. https://doi.org/10.2528/pierc19011106
[91] Banaei, H.A., Mehdizadeh, F.: Bandgap calculation of 2D hexagonal photonic crystal structures based on regression analysis. *J. Opt. Commun.* 34(4), 1–10. 2013. https://doi.org/10.1515/joc-2013-0033
[92] Boedecker, G., Henkel, C.: Validity of the quantum regression theorem for resonance fluorescence in a photonic crystal. *Annalen der Physik.* 524(12), 805–813. 2012. https://doi.org/10.1002/andp.201200135
[93] Bellingeri, M., Kriegel, I., Scotognella, F.: One dimensional disordered photonic structures characterized by uniform distributions of clusters. *Opt. Mat.* 39, 235–238. 2015. https://doi.org/10.1016/j.optmat.2014.11.033
[94] Lu, Z., Yin, Y.: Colloidal nanoparticle clusters: functional materials by design. *Chem. Soc. Rev.* 41(21), 6874–6887. 2012. https://doi.org/10.1039/c2cs35197h
[95] Saado, Y., Golosovksy, M.A., Davidov, D., Frenkel, A.: Tunable photonic band gap in self-assembled clusters of floating magnetic particles. *Phy. Rev. B.* 66(19), 2002. https://doi.org/10.1103/PhysRevB.66.195108
[96] Ma, W., Liu, Z., Kudyshev, Z.A., Boltasseva, A., Cai, W., Liu, Y.: Deep learning for the design of photonics structures. *Nat. Photonics.* 15, 77–90. 2021. https://doi.org/10.1038/s41566-020-0685-y

[97] Jiang, J., Chen, M., Fan, J.A.: Deep neural networks for the evaluation and design of photonic devices. *Nat. Rev. Mater.* 6, 679–700. 2021. https://doi.org/10.1038/s41578-020-00260-1

[98] Pfau, D., Spencer, J.S., Matthews, A.G., Foulkes, W.M.C.: Ab initio solution of the many-electron Schrödinger equation with deep neural networks. *Phys. Rev. Res.* 2(3), 033429. 2020. https://doi.org/10.1103/PhysRevResearch.2.033429

[99] Rodríguez-Esquerre, V. F.; Isídio-Lima, J. J.; Dourado-Sisnando, A.; Simões Silva, F. G.: Artificial neural networks for the chromatic dispersion prediction of photonic crystal fibers. *Microwave and Optical Technology Letters*, 55(9), 2179–2181. 2013. https://doi.org/10.1002/mop.27753

[100] Mirjalili, S.M., Mirjalili, S.Z.: Asymmetric oval-shaped-hole photonic crystal waveguide design by artificial intelligence optimizers. *IEEE J. Selec. Top. Quant. Electron.* 22(2), 258–264. 2015. https://doi.org/10.1109/JSTQE.2015.2469760

[101] Alagappan, G. Png, C.E.: Prediction of electromagnetic field patterns of optical waveguide using neural network. *Neural Comput. Appl.* 33, 2195–2206. 2021. https://doi.org/10.1007/s00521-020-05061-9

[102] Wiecha, P.R., Muskens, O.L.: Deep learning meets nanophotonics: A generalized accurate predictor for near fields and far fields of arbitrary 3D nanostructures. *Nano Lett.* 20, 329–338. 2019 https://doi.org/10.1021/acs.nanolett.9b03971

[103] Malkiel, I., Mrejen, M., Nagler, A., Arieli, U., Wolf, L., Suchowski, H.: Plasmonic nanostructure design and characterization via deep learning. *Light. Sci. Appl.* 7(1), 1–8. 2018. https://doi.org/10.1038/s41377-018-0060-7

[104] Inampudi, S., Mosallaei, H.: Neural network based design of metagratings. *Appl. Phys. Lett.* 112(24), 241102. 2018. https://doi.org/10.1063/1.5033327

[105] Pilozzi, L., Farrelly, F.A., Marcucci, G., Conti, C.: Machine learning inverse problem for topological photonics. *Commun. Phys.* 1(1), 1–7. 2018. https://doi.org/10.1038/s42005-018-0058-8

[106] Alagappan, G., Png, C.E.: Universal deep learning representation of effective refractive index for photonics channel waveguides. *J. Opt. Soc. Amer.* B. 36(10), 2636–2642. 2019. https://doi.org/10.1364/JOSAB.36.002636

[107] Nadell, C.C., Huang, B., Malof, J.M., Padilla, W.J.: Deep learning for accelerated all-dielectric metasurface design. *Opt. Express.* 27(20), 27523–27535. 2019. https://doi.org/10.1364/OE.27.027523

[108] Lio, G.E., Ferraro, A.: LIDAR and beam steering tailored by neuromorphic metasurfaces dipped in a tunable surrounding medium. *Photonics* 8(3), 65. 2021. https://doi.org/10.3390/photonics8030065

[109] Colburn, S., Majumdar, A.: Inverse design and flexible parameterization of meta-optics using algorithmic differentiation. *Commun. Phys.* 4(1), 1–11. 2021. https://doi.org/10.1038/s42005-021-00568-6

[110] Liu, Z., Feng, W., Long, Y., Guo, S., Liang, H., Qiu, Z., Fu, X., Li, J.: A Metasurface beam combiner based on the control of angular response. *Photonics* 8(11), 489. 2021. https://doi.org/10.3390/photonics8110489

[111] Alagappan, G., Png, C.E.: Modal classification in optical waveguides using deep learning. *J. Mod. Opt.* 66(5), 557–561. 2019. https://doi.org/10.1080/09500340.2018.1552331

[112] Alagappan, G., Png, C.E.: Deep learning models for effective refractive indices in silicon nitride waveguides. *J. Opt.* 21(3), 035801. 2019. https://doi.org/10.1088/2040-8986/ab00d5

[113] Colburn, S., Majumdar, A.: Inverse design and flexible parameterization of meta-optics using algorithmic differentiation. *Commun. Phys.* 4(1), 1–11. 2021. https://doi.org/10.1038/s42005-021-00568-6

[114] Minkov, M., Williamson, I.A., Andreani, L.C., Gerace, D., Lou, B., Song, A.Y., Hughes, T.W., Fan, S.: Inverse design of photonics crystals through automatic differentiation. *ACS Photonics* 7(7), 1729–1741. 2020. https://doi.org/10.1021/acsphotonics.0c00327

[115] Hughes, T.W., Minkov, M., Williamson, I.A., Fan, S.: Adjoint method and inverse design for nonlinear nanophotonics devices. *ACS Photonics* 5(12), 4781–4787. 2018. https://doi.org/10.1021/acsphotonics.8b01522

[116] Cao, Y., Li, S., Petzold, L., Serban, R.: Adjoint sensitivity analysis for differential-algebraic equations: The adjoint DAE system and its numerical solution. *SIAM J. Sci. Comput.* 24(3), 1076–1089. 2003 https://doi.org/10.1137/s1064827501380630

[117] Jiang, J., Sell, D., Hoyer, S., Hickey, J., Yang, J., Fan, J.A.: Free-form diffractive metagrating design based on generative adversarial networks. *ACS Nano.* 13(8), 8872–8878. 2019 https://doi.org/10.1021/acsnano.9b02371

[118] Jiang, J., Fan, J.A.: Global optimization of dielectric metasurfaces using a physics-driven neural network. *Nano Lett.* 19, 5366–5372. 2019. https://doi.org/10.1021/acs.nanolett.9b01857

[119] Lio, G.E., Ferraro, A., Ritacco, T., Aceti, D.M., De Luca, A., Giocondo, M., Caputo, R.: Leveraging on ENZ metamaterials to achieve 2D and 3D hyper-resolution in two-photon direct laser writing. *Adv. Mater.* 33(18), 2008644. 2021. https://doi.org/10.1002/adma.202008644

[120] Mellit, A., Eleuch, H., Benghanem, M., Elaoun, C., Pavan, A.M.: An adaptive model for predicting of global, direct and diffuse hourly solar irradiance. *Energy Convers. Manag.* 51(4), 771–782. 2010. https://doi.org/10.1016/j.enconman.2009.10.034

[121] Yang, Z., Hou, Y., Chen, Z., Zhang, L., Chen, J.: A multi-stage progressive learning strategy for covid-19 diagnosis using chest computed tomography with imbalanced data. In *Proceedings of the ICASSP 2021–2021 IEEE Int. Conf. Acoustics, Speech and Sig. Pro. (ICASSP)*, 8578–8582. 2021. https://doi.org/10.1109/ICASSP39728.2021.9414745

[122] Alagappan, G.; Png, C.E. Prediction of electromagnetic field patterns of optical waveguide using neural network. *Neural Comput. Appl.* 33, 2195–2206. 2021. https://doi.org/10.1007/s00521-020-05061-9

[123] Creswell, A., White, T., Dumoulin, V., Arulkumaran, K., Sengupta, B., Bharath, A. A.: Generative adversarial networks: An overview. *IEEE Signal Process. Mag.* 35(1), 53–65. 2018. https://doi.org/10.1109/MSP.2017.2765202

[124] Goodfellow, I., Pouget-Abadie, J., Mirza, M., Xu, B., Warde-Farley, D., Ozair, S., Courville, A., Bengio, Y.: Generative adversarial nets. *Adv. Neural Inf. Process. Syst.* 2, 2672–2680, 2014 https://doi.org/10.48550/arXiv.1406.2661

[125] He, K., Zhang, X., Ren, S., Sun, J.: Deep residual learning for image recognition. In *Proceedings of the IEEE Conference on Computer Vision and Pattern Recognition*, 770–778. 2016.

[126] Vaswani, A., Shazeer, N., Parmar, N., Uszkoreit, J., Jones, L., Gomez, A.N., Kaiser, Ł., Polosukhin, I.: Attention is all you need. *Comp. Lang.* 6000–6010. 2017. https://doi.org/10.48550/arXiv.1706.03762

[127] Hochreiter, S., Schmidhuber, J.: Long short-term memory. *Neural Comput.* 9(8), 1735–1780. 1997 https://doi.org/10.1162/neco.1997.9.8.1735

[128] Sherstinsky, A.: Fundamentals of recurrent neural network (RNN) and long short-term memory (LSTM) network. *Phys. D Non-linear Phenom.* 404, 132306 2020. https://doi.org/10.1016/j.physd.2019.132306

[129] Goodfellow, I., Pouget-Abadie, J., Mirza, M., Xu, B., Warde-Farley, D., Ozair, S., Courville, A., Bengio, Y.: Generative adversarial nets. *Adv. Neural Inf. Process. Syst.* 3(11), 2014. https://doi.org/10.48550/arXiv.1406.2661

[130] Kingma, D.P., Welling, M.: Auto-encoding variational bayes. *Stat. ML.* 1–9. 2013. https://doi.org/10.48550/arXiv.1312.6114

[131] Larochelle, H.; Murray, I. The neural autoregressive distribution estimator. In Proceedings of the Fourteenth International Conference on Artificial Intelligence and Statistics. *J. Mach. Reas.* 15, 29–37. 2011.

[132] Liu, Z., Zhu, D., Rodrigues, S.P., Lee, K.T., Cai, W.: Generative model for the inverse design of metasurfaces. *Nano Lett.* 18(10), 6570–6576. 2018. https://doi.org/10.1021/acs.nanolett.8b03171

[133] An, S., Zheng, B., Shalaginov, M.Y., Tang, H., Li, H., Zhou, L., Ding, J., Agarwal, A.M., Rivero-Baleine, C., Kang, M., Richardson, K. A., Gu, T., Hu, J., Fowler, C., Zhang, H.: Deep learning modeling approach for metasurfaces with high degrees of freedom. *Opt. Express.* 28(21), 31932–31942. 2020. https://doi.org/10.1364/OE.401960

[134] Ma, W., Cheng, F., Liu, Y.: Deep-learning-enabled on-demand design of chiral metamaterials. *ACS Nano* 12(6), 6326–6334. 2018. https://doi.org/10.1021/acsnano.8b03569

[135] Mao, S., Cheng, L., Zhao, C., Khan, F.N., Li, Q., Fu, H.Y.: Inverse design for silicon photonics: from iterative optimization algorithms to deep neural networks. *Appl. Sci.* 11(9), 1–25. 2021. https://doi.org/10.3390/app11093822

[136] Wu, X., Tian, J., Yang, R.: A type of all-optical logic gate based on graphene surface plasmon polaritons. *Opt. Commun.* 403, 185–192. 2017. https://doi.org/10.1016/j.optcom.2017.07.041

[137] Rezaei, M.H., Zarifkar, A.: Dielectric-loaded graphene-based plasmonic multilogic gate using a multimode interference splitter. *Appl. Opt.* 57(35), 10109–10116. 2018. https://doi.org/10.1364/AO.57.010109

[138] Rezaei, M.H., Zarifkar, A., Miri, M., Materials, O.: Ultra-compact electro-optical graphene-based plasmonic multi-logic gate with high extinction ratio. *Opt. Mater.* 84, 572–578. 2018. https://doi.org/10.1016/j.optmat.2018.07.043

[139] Sang, Y., Wu, Xiaojia, Raja, S.S., Wang, C.H., Li, H., Ding, Y., Liu, D., Zhou, J., Ahn H., Gwo, S., Shi, J.: Broadband multifunctional plasmonic logic gates. *Adv. Opt. Mater.* 6(13), 1701368 2018. https://doi.org/10.1002/adom.201701368

[140] Kita, S., Nozaki, K., Takata, K., Akihiko, S., Masaya, N.: Ultrashort low-loss Ψ gates for linear optical logic on Si photonics platform. *Commun Phys* 3, 33. 2020. https://doi.org/10.1038/s42005-020-0298-2

[141] Ying, Z. & Soref, R. Electro-optical logic using dual-nanobeam Mach-Zehnder interferometer switches. *Opt. Express* 29(9), 12801–12812. 2021.

[142] He, L., Zhang, W.X., Zhang, X.D.: Topological all-optical logic gates based on two-dimensional photonic crystals. *Opt Express.* 27(18), 25841–25859. 2019. https://doi.org/10.1364/OE.27.025841

[143] Qian, C., Lin, X., Lin, X., Xu, J., Sun, Y., Li, E., Zhang, B., Chen, H.: Performing optical logic operations by a diffractive neural network. *Light Sci Appl* 9, 59. 2020. https://doi.org/10.1038/s41377-020-0303-2

[144] Lin, C.E., Sun, C.P., Chen, C.C.: Utilizing optical neural network to establish high-performance OR and XOR logic gates. *Eng. App. Arti. Intel.* 119, 105788. 2023. https://doi.org/10.1016/j.engappai.2022.105788

[145] Zarei, S., Khavasi, A.; Realization of optical logic gates using on-chip difractive optical neural networks. *Sci Rep.* 12, 15747, 2022. https://doi.org/10.1038/s41598-022-19973-0

[146] Jandieri, V., Khomeriki, R., Onoprishvii, T., Werner, D.H., Berakdar, J., Erni, D.: Functional all-optical logic gates for true time-domain signal processing in nonlinear photonic crystal waveguides. *Opt. Express* 28, 18317–18331. 2020. https://doi.org/10.1364/OE.395015

[147] Yang, Z., Fu, Y., Yang, J., Hu, C., Zhang, J.: Spin-encoded subwavelength all-optical logic gates based on single-element optical slot nanoantennas. *Nanoscale* 10(9), 4523–4527. 2018. https://doi.org/10.1039.C7NR08871J–

[148] Liu, Y., Qin, F., Meng, Z.M., Zhou, F., Mao, Q.H., Li, Z.Y.: All-optical logic gates based on two-dimensional low-refractive-index nonlinear photonic crystal slabs. *Opt. Express.* 19(3), 1945–1953. 2011. https://doi.org/10.1364/oe.19.001945

[149] Liu, W., Yang, D., Shen, G., Tian, H., Ji, Y.: Design of ultra compact all-optical XOR, XNOR, NAND and OR gates using photonic crystal multi-mode interference waveguides. *Opt. Laser Technol.* 50, 55–64. 2013. https://doi.org/10.1016/j.optlastec.2012.12.030

[150] Wu, B., Ding, K., Chan, C.T., Chen, Y.: Machine prediction of topological transitions in photonic crystals. *Physical Review Applied* 14 (4), 044032. 2020. https://doi.org/10.1103/PhysRevApplied.14.044032

[151] Chen, Y., Liu, T., Wang, W.: Refractive index sensing performance analysis of photonic crystal Mach Zehnder interferometer based on BP neural network optimization. *Moder. Phys. Lett. B.* 29(10), 1550040. 2015. https://doi.org/10.1142/S0217984915500402

[152] El-Mosalmy, D.D., Hameed, M.F.O., Areed, N.F.F., Obayya, S.S.A.: Novel neural network based optimization approach for photonic devices. *Opt. Quant. Electron.* 46, 439–453. 2014. https://doi.org/10.1007/s11082-013-9869-8

[153] Hameed, M.F.O., Obayya, S.S.A., Al-Begain, K., Nasr, A.M., Abo el Maaty, M.I.: Accurate radial basis function based neural network approach for analysis of photonic crystal fibers. *Opt. Quant. Electron.* 40(11), 891–905. 2009. https://doi.org/10.1007/s11082-009-9290-5

[154] Burgess, A., Florescu M.: Modelling non-Markovian dynamics in photonic crystals with recurrent neural networks. *Optical Materials Express* 11(7), 2037–2048. 2021. https://doi.org/10.1364/OME.425263

[155] Alagappan, G., Png, C.E.: Meshless optical mode solving using scalable deep deconvolutional neural network. *Sci Rep.* 13(1078), 1–15. 2022. https://doi.org/10.1038/s41598-022-25613-4

[156] Anika, N.J., Mia, M.B.: Design and analysis of guided modes in photonic waveguides using optical neural network. *Optik.* 228, 1–11. 2020. https://doi.org/10.1016/j.ijleo.2020.165785

[157] Ferreira, A.S., Santos, C.H.S., Goncalves, M.S., Figueroa, H.E.H.: Towards an integrated evolutionary strategy and artificial neural network computational tool for designing photonic coupler devices. *Appl. Soft Comp.* 65(C), 1–11. 2017. https://doi.org/10.1016/j.asoc.2017.12.043

[158] Gabr, A.M., Fearherston, C., Zhang, C., Bonfil, C., Zhang, Q.J., Smy, T.J.: Design and optimization of optical passive elements using artificial neural networks. *J. Opt. Soc. America B.* 36(4), 2019. https://doi.org/10.1364/JOSAB.36.000999

[159] Liu, Z., Zhu, D., Raju, L., Cai, W.: Tackling photonic inverse design with machine learning. *Advanced Science* 8(5), 1–15. 2021 https://doi.org/10.1002/advs.202002923

[160] Singh, R., Agarwal, A., Anthony, B.W.: Mapping the design space of photonic topological states via deep learning. *Optics Exp.* 28(19), 1–9. 2020. 144. https://doi.org/10.1364/OE.398926

[161] Gan, L., Yu, F., Wang, Y., Wang, N., Zhu, X., Hu, L., Yu, C.: Dispersion-oriented inverse design of photonic-crystal fiber for four-wave mixing application. *Photonics.* 10(3), 294. 2023. https://doi.org/10.3390/photonics10030294

[162] Ren, Y., Zhang, L., Wang, W., Wang, X., Lei, Y., Xue, Y., Sun, X., Zhang, W.: Genetic-algorithm-based deep neural networks for highly efficient photonic device design. *Photon. Res.* 9(6), B247–B252. 2021. https://doi.org/10.1364/PRJ.416294

[163] Shen, L., Ye, Z., He, S.: Design of two-dimensional photonic crystals with large absolute band gaps using a genetic algorithm. *Phys. Rev. B*, 68(3), 1–5, 2003. 147. https://doi.org/10.1103/PhysRevB.68.035109

[164] Ghosh, A., Pal, A., Das, N.R.: An approach to design photonic crystal gas sensor using machine learning. *Optik*, 208, 1–13. 2019. https://doi.org/10.1016/j.ijleo.2019.163997

[165] Wittek, P. *Quantum Machine Learning: What Quantum Computing Means to Data Mining*; Academic Press, 2014.

[166] Sarma, S.D.; Deng, D.L.; Duan, L.M. Machine learning meets quantum physics. *Physics Today.* 72 (3), 48–54. 2019. https://doi.org/10.1063/PT.3.4164

[167] Kudyshev, Z.A.; Shalaev, V.M.; Boltasseva, A. Machine learning for integrated quantum photonics. *ACS Photonics.* 8, 34–46. 2020 https://doi.org/10.1021/acsphotonics.0c00960

[168] Haug, T., Mok, W.K., You, J.B., Zhang, W., Png, C.E., Kwek, L.C.: Classifying global state preparation via deep reinforcement learning. *Mach. Learn. Sci. Technol.* 2(1), 1–11. 2020. https://doi.org/10.1088/2632-2153/abc81f

[169] Wise, D.F., Morton, J.J., Dhomkar, S.: Using deep learning to understand and mitigate the qubit noise environment. *PRX Quantum.* 2, 1–15. 2021. 154. https://doi.org/10.1103/PRXQuantum.2.010316

[170] August, M., Ni, X.: Using recurrent neural networks to optimize dynamical decoupling for quantum memory. *Phys. Rev. A*, 95, 1–39. 2017. 155. https://doi.org/10.1103/PhysRevA.95.012335

[171] Kim, C., Park, K.D., Rhee, J.K.: Quantum Error Mitigation with Artificial Neural Network. *IEEE Access.* 8, 188853–188860. 2020. https://doi.org/10.1109/ACCESS.2020.3031607

[172] Cerezo, M., Arrasmith, A., Babbush, R., Benjamin, S.C., Endo, S., Fujii, K., McClean, J.R., Mitarai, K., Yuan, X., Cincio, L.: Variational quantum algorithms. *Nat. Rev. Phys.* 3, 625–644. 2021. https://doi.org/10.1038/s42254-021-00348-9

[173] Ewe, W.B., Koh, D.E., Goh, S.T., Chu, H.S., Png, C.E.: Variational quantum-based simulation of waveguide modes. *Quant. Phy.* 1–9. 2021. https://doi.org/10.1109/TMTT.2022.3151510

[174] You, J.B., Koh, D.E., Kong, J.F., Ding, W.J., Png, C.E., Wu, L. Exploring variational quantum eigen solver ansatzes for the long-range XY model. *Quant. Phy*. 1–9. 2021 https://doi.org/10.48550/arXiv.2109.00288

[175] Chen, S.Y.C., Yang, C.H.H., Qi, J., Chen, P.Y., Ma, X., Goan, H.S.: Variational quantum circuits for deep reinforcement learning. *IEEE Access*, 8, 141007–141024. 2020. https://doi.org/10.1109/ACCESS.2020.3010470

[176] Lockwood, O.; Si, M. Reinforcement learning with quantum variational circuit. In *Proceedings of the AAAI Conference on Artificial Intelligence and Interactive Digital Entertainment*, 35, 245–251. 2020 https://doi.org/10.1609/aiide.v16i1

Chapter 3

AI-integrated quantum networks for 6G

N. Alshaer and T. Ismail

3.1 INTRODUCTION

The rapid deployment of commercial 5G networks has led researchers to focus on the development of 6G networks. At the same time, the progress in quantum networks has opened up new possibilities for future 6G systems in terms of communication and computing shown in Figure 3.1. Quantum networks offer advanced capabilities, such as quantum teleportation and quantum computing, which could revolutionize the potential of 6G technology. By integrating quantum networks with 6G, the communication systems of the future could benefit from unparalleled levels of security, capacity, and capabilities. This integration could have a profound impact on various sectors, such as health care, autonomous vehicles, and the Internet of Things (IoT) [1, 2]. However, there are significant challenges to address, including the development of robust quantum technology, the establishment of quantum network infrastructure, and the integration of quantum algorithms into existing communication systems. Despite these obstacles, the convergence of quantum networks with 6G technology holds immense promise for transforming communication systems and enabling unprecedented levels of security, capacity, and capabilities.

3.2 QUANTUM SECURITY CONCEPTS IN EMERGING 6G

Emerging 6G is expected to revolutionize connectivity by offering ultra-high-speed and ultra-reliable networks. In order to realize these ambitious goals, integrating quantum technologies into the security aspects of 6G is gaining significant attention. Quantum security concepts involve leveraging the extraordinary properties of quantum mechanics to enhance the security measures employed in wireless communication systems. By harnessing

DOI: 10.1201/9781003591030-3

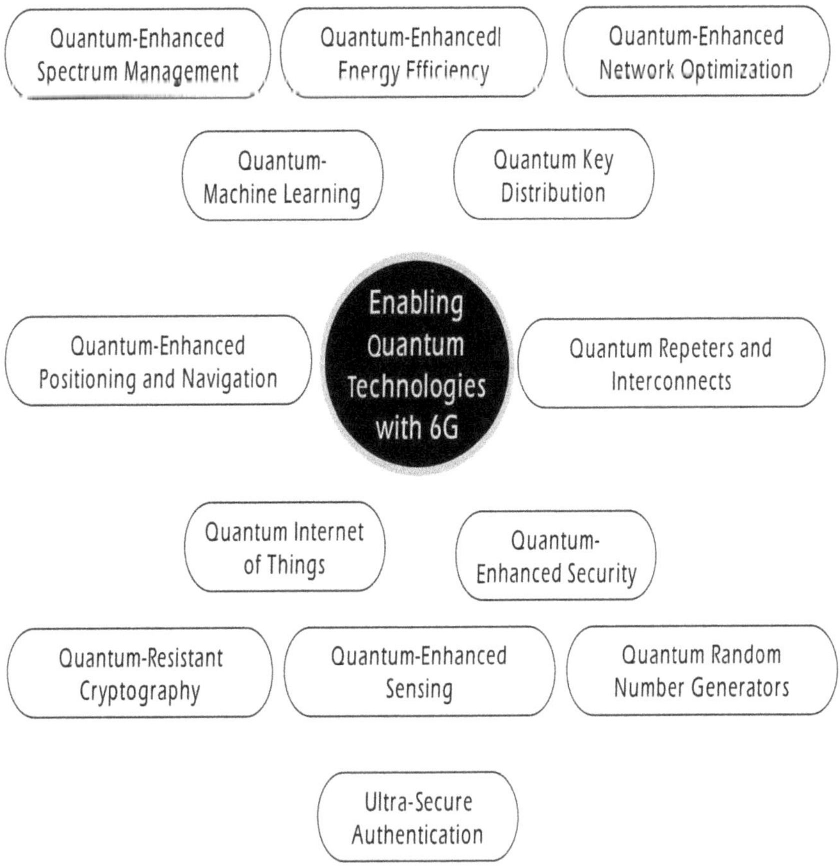

Figure 3.1 Enabling quantum technologies with 6G.

these quantum technologies, 6G networks can significantly improve their resistance against hacking, eavesdropping, and other forms of cyberattacks, thereby ensuring the confidentiality, authenticity, integrity, and availability of data. Quantum security concepts, therefore, have the potential to revolutionize the security landscape of 6G and enable the realization of unprecedented levels of secure and reliable wireless connectivity, as well as play a crucial role in achieving the following goals [3].

First, one of the key quantum security concepts in 6G is quantum key distribution (QKD). QKD utilizes the principles of quantum mechanics to securely exchange encryption keys between two communicating parties. It relies on the fundamental properties of quantum physics, such as the uncertainty principle and the no-cloning theorem, making it virtually impossible for an eavesdropper to intercept the key exchange without being detected. By integrating QKD into 6G networks, the communication channels can become inherently secure against any form of eavesdropping or hacking attempts [4]. Second, quantum-resistant cryptography is another important aspect of quantum security in 6G. With the advent of powerful quantum computers, most of the existing public-key encryption algorithms, such as Rivest, Shamir, Adleman (RSA), and Elliptic Curve Cryptography (ECC), will become vulnerable to attacks. Therefore, developing and deploying quantum-resistant cryptographic algorithms is crucial to ensuring the long-term security of 6G networks. These algorithms utilize mathematical problems that are believed to be hard for both classical and quantum computers, thereby maintaining the confidentiality and integrity of communication in the presence of advanced quantum adversaries [5].

The third quantum security concept in 6G involves quantum-based authentication and anti-counterfeiting techniques. Quantum mechanics offers unique properties, such as quantum entanglement and quantum superposition, that can be exploited to create unclonable and tamper-evident physical tokens. These tokens can be used for secure authentication of devices, individuals, or legitimate communication entities. By leveraging quantum authentication techniques, 6G networks can provide robust protection against identity theft, unauthorized access, and counterfeit attacks. Furthermore, quantum-based random number generation (QRNG) is another important concept in 6G quantum security [6]. Random numbers play a crucial role in various security protocols, such as encryption and key generation. However, classical computers can only generate pseudo-random numbers, which are deterministic and predictable. Quantum random number generators utilize the inherent randomness at the quantum level to provide truly random numbers that are fundamental for securing communication channels and cryptographic operations.

In addition to the aforementioned concepts, quantum security in 6G also involves quantum-resistant network security protocols and quantum-safe network management architectures [3]. These aim to develop secure communication protocols that can withstand quantum attacks and ensure the overall robustness of 6G networks. Moreover, quantum-safe network management architectures focus on designing and implementing mechanisms that can detect and respond to potential security threats posed by quantum computers, ensuring the continuity of secure communication in the presence of advanced quantum adversaries. Therefore, the emergence of 6G wireless communication systems presents a significant opportunity to integrate quantum security concepts to strengthen the security aspects of these networks. By leveraging quantum technologies such as QKD, quantum-resistant cryptography, quantum-based authentication, random number generation, network security protocols, and network management architectures, 6G networks can provide an unprecedented level of security and privacy for

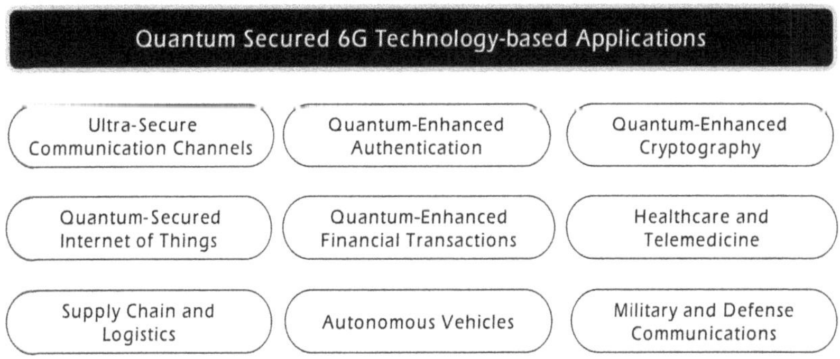

Figure 3.2 Quantum secured 6G technology-based applications.

users, ensuring the reliable and secure exchange of information in the future digital age (Figure 3.2).

3.3 INTEGRATED QUANTUM MACHINE LEARNING WITH 6G

Quantum machine learning, a field that combines quantum computing with traditional machine learning techniques, has the potential to greatly enhance the performance of 6G networks. This is crucial, as 6G networks aim to deliver unprecedented speeds, ultra-low latency, and reliable connectivity. Next, we will discuss some ways in which quantum machine learning can contribute to the improvement of 6G network performance [7].

First, quantum machine learning algorithms can help optimize resource allocation in 6G networks. With the massive increase in devices and data traffic expected in the coming years, intelligent resource management becomes imperative. Quantum machine learning algorithms can efficiently allocate network resources by considering complex factors and optimizing tasks such as spectrum allocation, power control, and bandwidth management. Second, quantum machine learning can enhance the security of 6G networks. As 6G networks embrace technologies, such as the IoT and artificial intelligence (AI), the surface for potential cyber threats increases significantly. Quantum machine learning can improve security solutions by leveraging quantum algorithms that are resistant to attacks from quantum computers, thereby safeguarding sensitive data and communications.

Moreover, quantum machine learning can assist in optimizing network routing and navigation in 6G networks. With the proliferation of diverse

IoT devices and emerging applications, 6G networks require coordination between different network elements. Quantum machine learning algorithms can analyze massive amounts of data and provide insights to optimize routing paths, reduce congestion, and improve overall network efficiency. Additionally, quantum machine learning can enhance anomaly detection and predictive maintenance in 6G networks. By analyzing vast amounts of data in real time, quantum algorithms can identify subtle network anomalies and predict potential failures before they occur. This can significantly reduce downtime, enhance network performance, and improve the quality of service provided by 6G networks.

Furthermore, quantum machine learning can facilitate intelligent edge computing in 6G networks. Edge computing, where data processing and storage occur closer to the end user, is crucial for enabling real-time applications and reducing latency. With the power of quantum machine learning, edge devices can efficiently analyze data, make intelligent decisions, and adapt to dynamic network conditions, leading to improved user experiences and efficient resource utilization, as well as enhancing the energy efficiency of 6G networks. This can be crucial, as 6G networks will likely involve billions of connected devices, and energy efficiency is of paramount importance.

In addition to the aforementioned, quantum machine learning can contribute to the development of advanced signal processing techniques in 6G networks. As 6G networks aim to deliver higher frequencies and bandwidths, traditional signal processing approaches face challenges due to inherent limitations. Quantum machine learning algorithms can help overcome these limitations and improve the efficiency of signal processing, enabling the use of high-frequency spectrum and enhancing the capacity and performance of 6G networks.

Another aspect where quantum machine learning can contribute to 6G network performance is in the field of data compression and encryption. Quantum algorithms can provide more effective compression techniques, reducing network bandwidth requirements while preserving data quality. Additionally, quantum encryption methods can provide stronger security measures to protect communications, making 6G networks more resilient against cyber threats. Therefore, quantum machine learning has the potential to revolutionize the performance of 6G networks in several ways. From optimizing resource allocation and improving security to enabling intelligent edge computing and efficient signal processing, the integration of quantum machine learning techniques can unlock the full potential of 6G networks, supporting the development of future technologies and applications [8] (Figure 3.3).

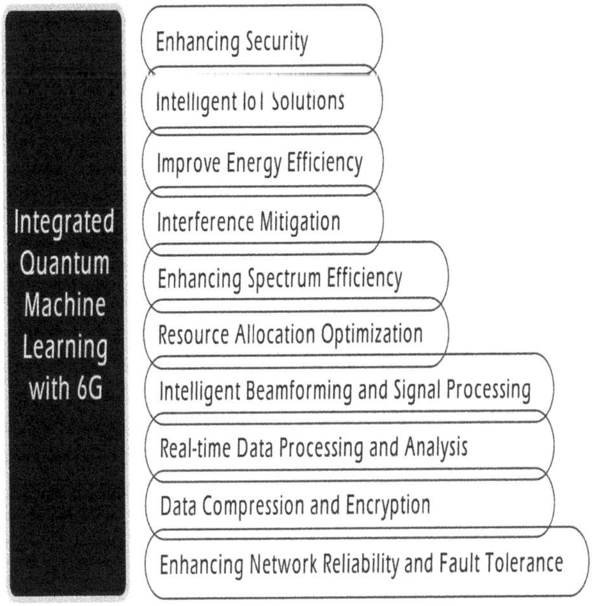

Figure 3.3 Integrated quantum machine learning with 6G.

3.4 AI-ENABLED QUANTUM NETWORK SCHEDULING IN 6G

AI-enabled quantum network scheduling in 6G networks is a groundbreaking technology that combines the power of AI and quantum computing to optimize the scheduling of network resources in the next generation of wireless networks. In 6G networks, the demand for high-speed, low-latency connections is expected to increase exponentially. Traditional scheduling algorithms may not be able to handle the complexity of these networks, resulting in poor network performance. However, with the integration of AI and quantum computing, network scheduling can be more efficient, leading to better overall network performance.

One of the key benefits of AI-enabled quantum network scheduling in 6G networks is the ability to adapt to changing network conditions in real time. By monitoring various network parameters, such as bandwidth, latency, and traffic load, AI algorithms can make intelligent scheduling decisions based on the current network state. Quantum computing further enhances this capability by performing complex computations in parallel, enabling faster decision-making. Moreover, AI-enabled quantum network scheduling can also address the security challenges associated with quantum networks. Quantum communication is highly secure, but it is vulnerable to attacks during the network scheduling process. With AI algorithms analyzing network traffic patterns in real time, it becomes easier to detect and prevent any malicious activities, ensuring the integrity and confidentiality of the network.

Furthermore, AI-enabled quantum network scheduling can revolutionize resource allocation in 6G networks. Traditional scheduling algorithms often allocate resources based on fixed priorities or predefined rules. However, with AI, dynamic resource allocation can be achieved by learning from past network performance data and predicting future demand. Quantum computing enables faster optimization of resource allocation, leading to improved network efficiency. Another advantage of AI-enabled quantum network scheduling is its ability to optimize energy consumption in 6G networks. By intelligently scheduling network resources and minimizing unnecessary operations, energy consumption can be significantly reduced, resulting in cost savings and more sustainable network infrastructure.

In addition to optimizing resource allocation, AI-enabled quantum network scheduling can also enhance network reliability. AI algorithms can analyze historical data and identify potential network failure points, allowing proactive measures to be taken to prevent disruptions. Quantum computing further enhances reliability by performing error correction and fault-tolerant computations.

The integration of AI and quantum computing in network scheduling also opens up new possibilities for advanced services in 6G networks. For example, AI algorithms can analyze user preferences and behavior to provide personalized services, such as intelligent content recommendations or adaptive quality of service. Furthermore, AI-enabled quantum network scheduling can facilitate the seamless integration of heterogeneous networks in 6G. By optimizing the utilization of network resources across different technologies, such as satellite, terrestrial, and underwater networks, AI algorithms can provide a unified and seamless user experience, regardless of the network type or location.

In addition to the aforementioned, AI-enabled quantum network scheduling can pave the way for autonomous network management in 6G networks. With AI algorithms continuously monitoring and optimizing network resources, network operators can focus on higher-level tasks, such as strategic planning and business development, while the network operates autonomously. Therefore, AI-enabled quantum network scheduling in 6G networks is a game-changing technology that combines the power of AI and quantum computing to optimize network performance, enhance security, improve resource allocation, reduce energy consumption, enhance reliability, enable advanced services, facilitate network integration, and enable autonomous network management. This technology has the potential to revolutionize the future of wireless networks and unlock unprecedented opportunities for innovation and growth (Figure 3.4).

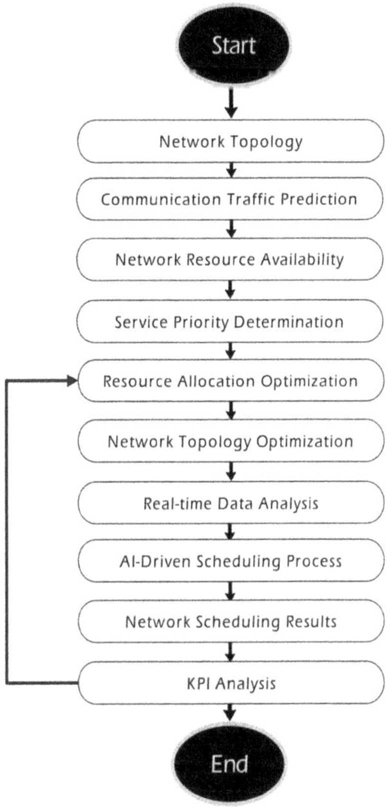

Figure 3.4 AI-enabled quantum network scheduling in 6G.

3.5 QUANTUM NEURAL NETWORKS FOR 6G COMMUNICATION

Quantum neural networks (QNNs) are a new paradigm that combines the principles of quantum computing with artificial neural networks (ANNs). With the advent of 6G networks on the horizon, there is a growing demand for more powerful and efficient computational tools. QNNs have the potential to revolutionize the field of machine learning by leveraging the unique properties that quantum systems offer. Unlike classical neural networks, QNNs can utilize the principles of superposition and entanglement, allowing for enhanced computational capabilities and improved learning algorithms [9]. The integration of QNNs into 6G networks holds great promise for various applications. For instance, QNNs can be employed in intelligent routing algorithms, enabling faster and more efficient data transfers. Additionally, QNNs can enhance security protocols, as they offer better resistance against quantum attacks. With the increasing complexity of network infrastructures, the use of QNNs can help optimize network traffic, reduce latency, and improve overall network performance [10].

One of the key advantages of QNNs in 6G networks is their ability to process and analyze large amounts of data in a parallel manner. Quantum superposition allows for the simultaneous evaluation of multiple states, leading to faster computations and enhanced pattern recognition capabilities. This capability will be particularly beneficial in areas such as image and speech recognition, where large datasets need to be processed in real time. Another notable feature of QNNs in 6G networks is their ability to take advantage of quantum entanglement. Entanglement allows for the correlation of states across different qubits, even when physically separated. This property opens possibilities for distributed computation, where data can be processed on separate quantum nodes but still maintain a correlated and coherent representation. This distributed processing capability can greatly enhance the scalability and performance of 6G networks [11].

Quantum support vector machines (SVMs) and ANNs are emerging technologies in the field of QNNs that offer significant improvements over classical SVMs and ANNs. These quantum algorithms leverage the principles of quantum mechanics, such as superposition and entanglement, to enhance their performance and offer advantages in solving complex classification and pattern recognition problems. Quantum SVMs utilize qubits, the basic unit of quantum information, to encode data and perform computations. By taking advantage of quantum parallelism, SVMs can evaluate multiple data points simultaneously, leading to faster classification and improved accuracy. Moreover, quantum SVMs are also effective in handling high-dimensional feature spaces, whereas classical SVMs may struggle due to computational limitations. Quantum ANNs, on the other hand, harness quantum circuits to encode and process data in a way that allows them to capture complex relationships and patterns within the data. These networks use qubits for efficient representation of input data and quantum gates for performing computations. The ability of quantum ANNs to simultaneously process a large number of input patterns can expedite training and inference processes, leading to speed-up in solving complex problems [8].

The integration of quantum SVMs and ANNs with 6G, the next generation of wireless communication technology, offers numerous possibilities for enhancing wireless network performance. By leveraging the capabilities of quantum computing, 6G networks can benefit from more efficient resource allocation, improved spectrum sensing, and enhanced security mechanisms. For example, quantum SVMs can aid in optimizing radio resource allocation by efficiently handling the complexities posed by a large number of devices and rapidly changing network conditions. Another area where quantum SVMs and ANNs can contribute to 6G is in spectrum sensing and utilization. Quantum algorithms can assist in detecting and utilizing idle spectrum bands, enabling more efficient spectrum sharing among different network operators and devices. This can lead to improved spectrum utilization, reduced interference, and better quality of service for users.

Therefore, the integration of QNNs into 6G networks holds tremendous potential for advancing computational capabilities and revolutionizing machine learning. The unique properties offered by quantum systems, such as superposition and entanglement, can significantly enhance the performance and efficiency of network infrastructures. However, there are still several technological, algorithmic, and ethical challenges that need to be addressed before the widespread adoption of QNNs in 6G networks. Continued research and collaboration between quantum physicists, computer scientists, and network engineers are essential for unlocking the full potential of QNNs in 6G networks.

3.6 QUANTUM SWARM INTELLIGENCE FOR 6G

Swarm intelligence is inspired by the collective behavior of social insect colonies, where simple individuals collectively solve complex problems. Quantum swarm intelligence is a concept that combines quantum computing with swarm intelligence techniques to enhance the efficiency, scalability, and security of wireless communication systems [12]. With 6G networks expected to support unprecedented data rates, massive connectivity, and ultra-low latency, quantum swarm intelligence offers unique solutions that can revolutionize the way future wireless networks are designed and operated [13].

One of the key advantages of using quantum swarm intelligence in 6G applications is its potential to significantly improve the efficiency of spectrum allocation. By leveraging the principles of swarm intelligence, multiple devices can cooperate and allocate available spectrum resources dynamically. Quantum computation can further enhance this process by efficiently solving complex optimization problems, enabling the allocation of spectrum resources in real time [14]. This results in increased spectral efficiency and better utilization of the finite radio spectrum, crucial factors for 6G networks. Furthermore, the scalability of quantum swarm intelligence makes it an ideal candidate for accommodating the massive number of devices expected to be connected to 6G networks. With the proliferation of IoT devices and the integration of various smart systems, future networks will need to handle a vast number of connected devices. Quantum swarm intelligence can enable efficient device coordination, resource allocation, and network management, allowing 6G networks to support the massive connectivity requirements of the future.

In addition to improved efficiency and scalability, quantum swarm intelligence also has the potential to enhance the security of 6G applications. Quantum computation offers superior encryption and data security techniques, such as QKD, which can ensure secure communication in the network. Moreover, swarm intelligence enables the detection and mitigation of malicious activities by leveraging collective intelligence and distributed decision-making, providing an additional layer of security in 6G networks.

Another area where quantum swarm intelligence can make a significant impact in 6G applications is network planning and optimization. Quantum computation techniques can be employed to efficiently solve optimization problems related to coverage, capacity, and quality of service. Additionally, swarm intelligence algorithms can aid in self-organizing network configurations and adaptive network optimization, thereby improving network performance and user experience.

The use of quantum swarm intelligence can also play a pivotal role in intelligent resource management in 6G networks. By combining the inherent intelligence of swarm algorithms with quantum computation, network resources can be allocated adaptively and dynamically depending on the varying requirements of applications and users. This not only enhances resource utilization but also enables efficient energy management, a critical aspect in future networks aiming for sustainability. Moreover, the integration of quantum swarm intelligence with 6G networks can lead to breakthroughs in autonomous communication and decision-making. Swarm intelligence algorithms can enable decentralized decision-making and distributed task allocation among devices, allowing for adaptive and self-organizing networks. Quantum computation techniques further enhance these capabilities by efficiently processing massive amounts of data and enabling real-time intelligent decision-making.

In addition to the aforementioned, quantum swarm intelligence can also provide innovative solutions for emerging applications in 6G networks, such as virtual reality (VR), augmented reality (AR), and haptic communication. The combination of quantum computation techniques and swarm algorithms can enable efficient content distribution, low-latency communication, and immersive user experiences, taking 6G applications to new levels of performance and usability. Therefore, quantum swarm intelligence holds significant promise for addressing the challenges of 6G applications. Its ability to enhance efficiency, scalability, security, network planning, resource management, autonomous communication, and support emerging applications makes it a compelling approach for future wireless networks. As researchers and industry experts work toward developing 6G technologies, quantum swarm intelligence will likely continue to gain attention as a key enabler for achieving the ambitious goals of next-generation communication systems.

3.7 AI-ENABLED QUANTUM DATA ANALYTICS FOR 6G NETWORKS AND APPLICATIONS

AI-enabled quantum data analytics for 6G networks and applications represent the next frontier in telecommunications and data analysis. As we look toward the development of 6G networks and the vast amount of data they will generate, traditional analytics methods may no longer be sufficient to handle the complexity and volume of information [15]. This is where AI-powered quantum data analytics steps in, utilizing the computational power of quantum computers and the intelligence of artificial intelligence algorithms to unlock new insights and optimize the performance of 6G networks.

One of the key advantages of AI-enabled quantum data analytics is its ability to handle massive amounts of data. As 6G networks start operating, they will generate an unprecedented amount of information from various sources, including IoT devices, sensors, and social media. Traditional analytics methods struggle to process such massive data, but quantum computers can handle this task due to their parallel processing capabilities. Another benefit of AI-enabled quantum data analytics is its ability to identify complex patterns and correlations within the data. Quantum algorithms can efficiently search for hidden relationships that may not be apparent to traditional analytics methods, leading to more accurate predictions and insights. This capability will be crucial for optimizing 6G network performance, as it allows for proactive maintenance and identification of potential bottlenecks [16].

AI-powered quantum data analytics can also improve network security for 6G networks. Quantum algorithms can analyze massive amounts of network data and identify potential security breaches in real time. With the constantly evolving nature of cyber threats, this proactive approach to security is essential to protect the networks and the sensitive information they carry. Moreover, AI-enabled quantum data analytics can enhance resource allocation and optimization in 6G networks. By analyzing real-time data on network traffic, quantum algorithms can dynamically allocate resources based on demand, thereby maximizing network efficiency. This real-time optimization can lead to improved network bandwidth, reduced latency, and better overall user experience.

Furthermore, AI-powered quantum data analytics can help in the development of intelligent network management systems. By combining quantum computing capabilities with AI algorithms, network managers can make more informed decisions on network planning, routing, and capacity management. These intelligent systems can continuously learn from network data to optimize performance and adapt to changing network conditions. Additionally, AI-enabled quantum data analytics can facilitate the implementation of advanced beamforming techniques in 6G networks. Quantum algorithms can analyze vast amounts of data to calculate and optimize the beamforming parameters in real time. This will enable the efficient use of spectrum resources, improved coverage, and higher data rates in 6G networks [17].

Moreover, AI-powered quantum data analytics can assist in the development of intelligent network traffic prediction models. By analyzing historical traffic data along with other relevant factors, these models can accurately predict future traffic patterns. This information is invaluable for network planning and resource allocation, ensuring that the 6G networks are ready to handle anticipated peak loads. Furthermore, AI-enabled quantum data analytics can facilitate the implementation of advanced anomaly detection systems in 6G networks. Quantum algorithms can analyze network data and identify abnormal patterns or behaviors, indicating potential faults, cyberattacks, or network congestion. Early detection of these anomalies allows for prompt action to mitigate their impact and ensure network stability. Therefore, AI-enabled quantum data analytics represents a revolutionary approach to handling the massive amounts of data generated by 6G networks. The combination of quantum computing capabilities and AI algorithms allows for the efficient processing of data, identification of complex patterns, enhanced network optimization, improved security, and development of intelligent network management systems. As we move toward the era of 6G, AI-powered quantum data analytics will play a crucial role in ensuring the performance, reliability, and security of these advanced networks (Figure 3.5).

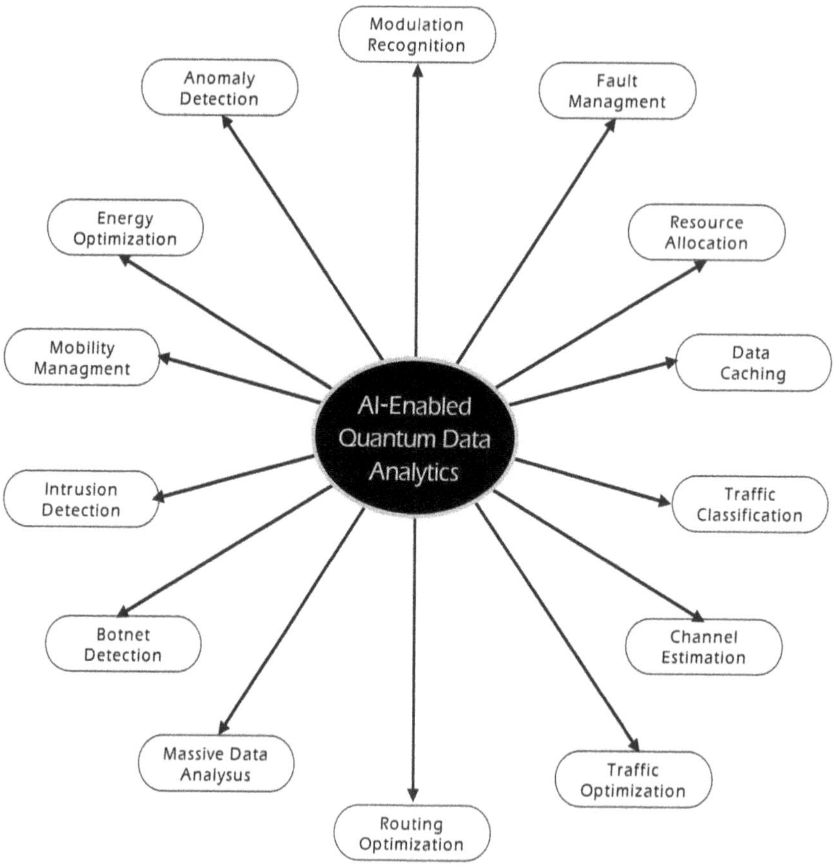

Figure 3.5 AI-enable quantum data analysis in 6G networks and applications.

3.8 QUANTUM EDGE AND CLOUD COMPUTING WITH EMPOWERED AI FOR 6G

Quantum edge and cloud computing with empowered AI for 6G applications is an emerging field that combines advanced technologies to drive the next generation of wireless communication and computing. This integration promises to revolutionize various sectors, including telecommunications, healthcare, transportation, and energy. Also, it holds immense promise for transforming our digital landscape. From enhanced network capabilities and secure communication to intelligent services and improved efficiencies across industries, this powerful convergence of technologies will pave the way for a future where data is processed, analyzed, and transmitted faster and more efficiently than ever before [18].

Quantum edge computing brings the power of quantum computing to the edge of the network, enabling real-time data processing and analysis at the source [13]. By eliminating the need to transmit data to a centralized cloud server, quantum edge computing significantly reduces latency and optimizes network bandwidth, making it ideal for time-sensitive 6G applications [19]. Cloud computing plays a vital role in this ecosystem by providing a centralized platform for managing and storing vast amounts of data generated by edge devices. By utilizing cloud infrastructure, quantum edge computing can securely process and analyze data from various edge devices, such as smartphones, IoT sensors, and autonomous vehicles. This decentralized approach ensures real-time data analysis and storage, enabling seamless integration of 6G applications. Empowered AI is another crucial component of this technology stack. By harnessing the power of machine learning, empowered AI enables cognitive computing in real time. This capability allows quantum edge computing systems to make autonomous decisions, adapt to changing environments, and deliver personalized services based on user preferences [20].

One of the key applications of quantum edge and cloud computing with empowered AI is in the telecommunications industry. With the advent of 6G, there will be an exponential increase in data traffic, which requires ultra-low latency and high bandwidth networks. Quantum edge computing can offload complex computational tasks from mobile devices to edge servers, significantly reducing latency and improving user experience. In the healthcare sector, quantum edge computing with empowered AI can revolutionize personalized medicine and remote patient monitoring. By leveraging advanced analytics and machine learning algorithms, healthcare providers can deliver accurate diagnoses, monitor patients in real time, and develop personalized treatments based on individual genetic profiles. Transportation is another industry that can benefit from this integration. With autonomous vehicles becoming more prevalent, quantum edge computing can enable real-time data processing for navigation, object recognition, and decision-making. This technology can also enhance vehicle-to-vehicle communication, ensuring safe and efficient transportation systems.

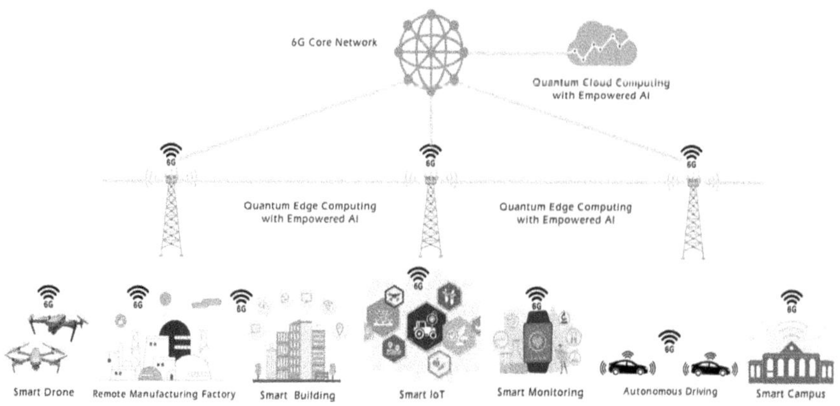

Figure 3.6 Quantum edge and cloud computing with empowered AI.

The energy sector can benefit from the quantum edge and cloud computing with empowered AI by optimizing power generation and distribution. By analyzing real-time data from smart grids, energy providers can improve energy efficiency, reduce carbon footprint, and predict future energy demands more accurately. The security of 6G applications is of paramount importance. Quantum cryptographic techniques, which use the principles of quantum mechanics, can provide unbreakable encryption, ensuring the confidentiality and integrity of sensitive data transmitted over the network. The deployment of quantum edge computing combined with empowered AI enhances the security measures of 6G applications, protecting them against emerging threats.

The possibilities of quantum edge and cloud computing with empowered AI for 6G applications are not limited to these industries alone. It can also be used in smart cities, industrial automation, financial services, AR, VR and many other sectors. As these technologies continue to evolve and mature, we can expect a seamless integration of quantum computing, edge computing, cloud computing, and AI to unlock innovative solutions and drive economic growth (Figure 3.6).

3.9 CONCLUSION AND CHALLENGES

In conclusion, the integration of AI in quantum networks for 6G holds massive potential to revolutionize various industries and improve overall connectivity. The combination of AI and quantum computing capabilities can enable faster data processing, more efficient networking, and enhanced security in 6G networks. By employing AI algorithms and machine learning techniques, quantum networks can adapt and optimize their performance in real time. This ensures efficient utilization of network resources and effective

management of network traffic, leading to increased reliability, reduced latency, improved security, enhanced data processing, allocated resources efficiently, and improved overall performance in 6G networks.

However, several challenges need to be addressed for the successful implementation of AI-integrated quantum networks in 6G. The challenge in building large-scale quantum networks lies in creating a robust infrastructure capable of accommodating a vast number of interconnected quantum devices while ensuring interoperability of hardware, software, and algorithms to enable seamless communication. Quantum network management and control are also significant challenges, as sophisticated control mechanisms are needed to maintain delicate quantum states and accurately route quantum information. The development of quantum repeaters is necessary to overcome signal loss in optical fibers and enable reliable quantum communication over long distances. Successfully integrating AI with quantum networks in the context of 6G requires designing frameworks that allow for seamless interaction between AI algorithms and quantum devices to enhance decision-making and optimization processes. Quantum error correction is crucial to mitigate the effects of noise and decoherence, ensuring accuracy and reliability. The establishment of standards and regulations for AI-integrated quantum networks in 6G is critical for global interoperability, privacy protection, and data security. Finally, developing hybrid networks that integrate classical and quantum systems requires seamless integration, compatibility, and efficient resource allocation to leverage the strengths of both technologies.

REFERENCES

[1] C. Wang and A. Rahman, "Quantum-enabled 6G wireless networks: opportunities and challenges," *IEEE Wireless Communications*, vol. 29, no. 1, pp. 58–69, February 2022, doi: 10.1109/MWC.006.00340

[2] M. Z. Ali, A. Abohmra, M. Usman, A. Zahid, H. Heidari, M. A. Imran, Q. H. Abbasi, "Quantum for 6G communication: A perspective," *IET Quantum Communication*, pp. 1–13, May 2023, doi: 10.1049/qtc2.12060

[3] S. H. A. Kazmi, R. Hassan, F. Qamar, K. Nisar, A. A. A. Ibrahim, "Security concepts in emerging 6G communication: threats, countermeasures, authentication techniques and research directions", *Symmetry*, vol. 15, no. 6, pp. 1–33, May 2023, doi: 10.3390/sym15061147

[4] M. H. Adnan, Z. Z. Ahmad, N. Z. Harun, "Quantum key distribution for 5G networks: a review, state of art and future directions," *Future Internet*, vol. 14, no. 73, pp. 1–28, February 2022, doi: 10.3390/fi14030073

[5] M. Allende, D. L. León, S. Cerón, et al. "Quantum-resistance in blockchain networks," *Scientific Report*, vol. 13, no. 5664, pp. 1–23, April 2023, doi: 10.1038/s41598-023-32701-6

[6] M. Siswanto and B. Rudiyanto, "Designing of quantum random number generator (QRNG) for security application," *2017 3rd International Conference*

on Science in Information Technology (ICSITech), Bandung, Indonesia, 2017, pp. 273–277, doi: 10.1109/ICSITech.2017.8257124

[7] S. J. Nawaz, S. K. Sharma, S. Wyne, M. N. Patwary and M. Asaduzzaman, "Quantum machine learning for 6G communication networks: state-of-the-art and vision for the future," *IEEE Access*, vol. 7, pp. 46317–46350, 2019, doi: 10.1109/ACCESS.2019.2909490

[8] F. Vista, V. Musa, G. Piro, L. A. Grieco and G. Boggia, "Network intelligence with quantum computing in 6G and B6G: design principles and future directions," *2021 IEEE Globecom Workshops (GC Wkshps)*, Madrid, Spain, 2021, pp. 1–6, doi: 10.1109/GCWkshps52748.2021.9682045

[9] Y. Kwak, W. J. Yun, S. Jung and J. Kim, "Quantum neural networks: concepts, applications, and challenges," *2021 Twelfth International Conference on Ubiquitous and Future Networks (ICUFN), Jeju Island, Korea, Republic of*, 2021, pp. 413–416, doi: 10.1109/ICUFN49451.2021.9528698

[10] S. Riyaz, K. Sankhe, S. Ioannidis, and K. Chowdhury, "Deep learning convolutional neural networks for radio identification," *IEEE Communications Magazine*, vol. 56, no. 9, pp. 146–152, Sep. 2018

[11] M. Chehimi and W. Saad, "Quantum federated learning with quantum data," *ICASSP 2022 - 2022 IEEE International Conference on Acoustics, Speech and Signal Processing (ICASSP)*, Singapore, Singapore, 2022, pp. 8617–8621, doi: 10.1109/ICASSP43922.2022.9746622

[12] Q. Pham, D. C. Nguyen, S. Mirjalili, D. T. Hoang, D. N. Nguyen, P. N. Pathirana and W. Hwang, "Swarm intelligence for next-generation networks: Recent advances and applications," *Journal of Network and Computer Applications*, vol. 191, Oct. 2022, doi: 10.1016/j.jnca.2021.103141

[13] T. Q. Duong, L. D. Nguyen, B. Narottama, J. A. Ansere, D. V. Huynh and H. Shin, "Quantum-inspired real-time optimization for 6G networks: opportunities, challenges, and the road ahead," *IEEE Open Journal of the Communications Society*, vol. 3, pp. 1347–1359, Aug. 2022, doi: 10.1109/OJCOMS.2022.3195219

[14] J. Zhou, H. Qin, Y. Zhang, R. Yang, Y. Liu, & C. Li, "Binary quantum elite particle swarm optimization algorithm for spectrum allocation in cognitive wireless medical sensor network," *Journal of Physics: Conference Series*, vol. 1924, no. 1, pp. 012030.1–012030.7, May 2021, doi:10.1088/1742-6596/1924/1/012030

[15] T. B. Ahammed, R. Patgiri, & S. Nayak, " A vision on the artificial intelligence for 6G communication," *ICT Express*, vol. 9, no. 2, pp. 197–210, April 2023, doi: 10.1016/j.icte.2022.05.005

[16] C. Fiandrino, G. Attanasio, M. Fiore, & J. Widmer, "Toward native explainable and robust AI in 6G networks: Current state, challenges and road ahead," *Computer Communications*, vol. 193, pp. 47–52, Sept. 2022, doi: 10.1016/j.comcom.2022.06.036

[17] A. M. Al-Ansi, & A. Al-Ansi, "An overview of Artificial Intelligence (AI) in 6G: types, advantages, challenges and recent applications," *Buletin Ilmiah Sarjana Teknik Elektro*, vol. 5, no.1, pp. 67–75, Mar. 2023, doi: 10.12928/biste.v5i1.7603

[18] A. Bandi, "A review towards AI empowered 6G communication requirements, applications, and technologies in mobile edge computing," In *6th International Conference on Computing Methodologies and Communication (ICCMC)*, Erode, India, Mar. 2022, pp. 12–17, doi: 10.1109/ICCMC53470.2022

[19] K. B. Letaief, Y. Shi, J. Lu, & J. Lu, "Edge artificial intelligence for 6G: Vision, enabling technologies, and applications," *IEEE Journal on Selected Areas in Communications*, vol. 40, no.1, pp. 5–36, Jan. 2021, doi: 10.1109/JSAC.2021.3126076

[20] M. Z. Asghar, S. A. Memon, & J. Hämäläinen, "Evolution of wireless communication to 6g: Potential applications and research directions," *Sustainability*, vol. 14, no.10, pp. 1–26, May 2022, doi: 10.3390/su14106356

Chapter 4

Development in AI for biophotonics applications

Than Singh Saini

4.1 INTRODUCTION

The term "biophotonics" refers to the understanding of the interaction of biological samples with the photons or light. Biophotonics utilizes light across the entire spectrum, ranging from the ultraviolet to the visible, infrared, and terahertz domains. In order to study the important characteristics, biological sample may also include any food related items. Since the development of the optical microscope in the 17th century, light-based technologies have greatly influenced the life sciences and medicine. The biological sample can be divided based on the dimensions of interest into single cells, organs, tissues, and whole body. In the 1960s, the invention of the laser boosted several optical techniques, including spectroscopy and imaging. Basically, biophotonics is a field of science that deals with optical processes in biological systems. The 2014 Nobel Prize in chemistry was awarded for the development of the ultra-high resolution microscope, which is able to visualize cellular structures even smaller than 20 nm. Such inventions significantly benefit the field of biomedicine.

Biophotonics has been a rapidly growing field with great potential to become a part of clinical trials. After the invention of lasers in the 1960s, numerous optical techniques, including imaging and spectroscopy, have advanced exponentially. Such advancements have established corresponding applications in the field of biophotonics.

Further, in the 20th century, the observation of regularities in the atomic spectra has played a vital role in the verification and development of quantum physics. The spontaneous identification of consistencies present in the data has played a crucial role in the discovery of the fundamental laws of the sciences. This prerequisite of recognizing the patterns in large-size datasets needs automatic methods of analysis, which are provided by machine learning. Machine learning is a subfield of artificial intelligence (AI), which comprises the learning of computational algorithms that improve automatically without being explicitly programmed [1].

DOI: 10.1201/9781003591030-4

The necessity for adaptability and flexibility of various sorts of data has been attained by using machine learning–based techniques. The capability for automated learning without explicitly outlining the rules provided these techniques with an edge against any other techniques. Initial techniques for the analysis were applied by reviewing multivariate techniques, including principal component analysis or linear discriminant analysis. Nonetheless, these techniques being linear, present disadvantage when applied to non-linearly categorized datasets, which is a common trend in the biomedical data. Artificial neural networks (ANNs) and deep learning offer modern advancement in machine learning. Because of their ease of applicability to different types of data, ANNs and deep learning have been used in abundant applications ranging from the bioinformatics to photonics fields [2, 3].

This chapter is devoted to exploring the applications of machine learning and finding the solutions that may enhance the working of optical instruments. The important application of machine learning in label-free characterization of immune cells is mainly discussed. The recent developments in the field of biophotonics have been included since biophotonics has the potential to offer various applications, including the detection of chemical and biological molecules, microorganisms, and biomedical imaging. In such fields, we have a large amount of generated informative data of high complexity. AI is a disruptive technology with applications that now pervade every field of our knowledge. The various components of AI, including reinforcement learning, machine learning, and deep learning, are capable not only of finding hidden patterns in complex data but also stimulating new theoretical and experimental interpretations. Further, this chapter includes the purpose of exploring the crossroads between the AI and biophotonics disciplines, with the aim of creating a knowledge path for the development of AI applications in the field of biophotonics.

4.2 METHODS IN MACHINE LEARNING

Linear methods of machine learning include linear discriminant analysis (LDA), k-means clustering, principal component analysis (PCA), and support vector machines. The non-linear methods include Adaboost, T-distributed stochastic neighbor embedding (t-SNE) and ANNs.

4.2.1 Linear discriminant analysis

LDA is a linear method of machine learning based on finding the linear combination of characteristic features for the classification of input datasets. This algorithm is generally useful for reducing the dimensionality of any data and then classifying it linearly. Basically, the LDA and PCA are linear methods and are closely related. LDA and PCA both work on the principal of linear combination of variables to reduce the dimensionality of the dataset. The difference between LDA and PCA is that PCA identifies the degree of variance representing the differences between the classes, while LDA models the similarities of various components in terms of features or patterns. LDA is useful where the prior knowledge of classes in the fundamental data is known, while PCA can be executed without any previous knowledge.

4.2.2 K-means algorithm

The k-means algorithm is based on vector quantization. This is a linear method that aims to separate input datasets into k discrete nonoverlapping clusters, providing each data point fits into a given cluster pertaining to the cluster's mean. The k-means algorithm follows the method of expectation-maximization, in which the data points $\left(x_i \epsilon \left\{x_1, x_2, \ldots, x_n\right\}\right)$ are assigned by keeping the sum of the squared distance to the cluster's centroid to be at the minimum. First, the number of clusters k is specified, and then by randomly shuffling the data and choosing the random k data points, centroids for each cluster are initialized. Such action is iterated until we reach the situation where no change in the assignment of centroid is obtained. For its broad applicability, the k-means algorithm has initiate numerous applications. It has been applied to probabilistic-shaped 64 quadrature amplitude modulation coherent optical communication systems to locate decision points more precisely, as a benchmark for digital staining of optical coherence tomography images and for optic disc segmentation [4, 5].

4.2.3 Principal component analysis

PCA is a highly notable statistical procedure for identifying the presence of the patterns in any sort of dataset. This method uses an orthogonal transformation to convert a set of correlated variables into the eigenvalue eigenvector problem. The eigenvectors are, in general, called principal components. The number of principal components is generally less than the number of original variables. PCA is implemented by collecting a dataset in the "n" dimensions. The mean is subtracted from the collected data in order to eliminate the effects of the fluctuations in intensity. Following which, a covariance matrix is calculated, and the eigenvectors and eigenvalues are extracted from this matrix. The first principal component describes the maximal variation, which is directly associated with the eigenvector with the largest eigenvalue.

4.2.4 Support vector machines

The support vector machines are the supervised learning models of machine leaning. These are generally executed for the regression analysis and

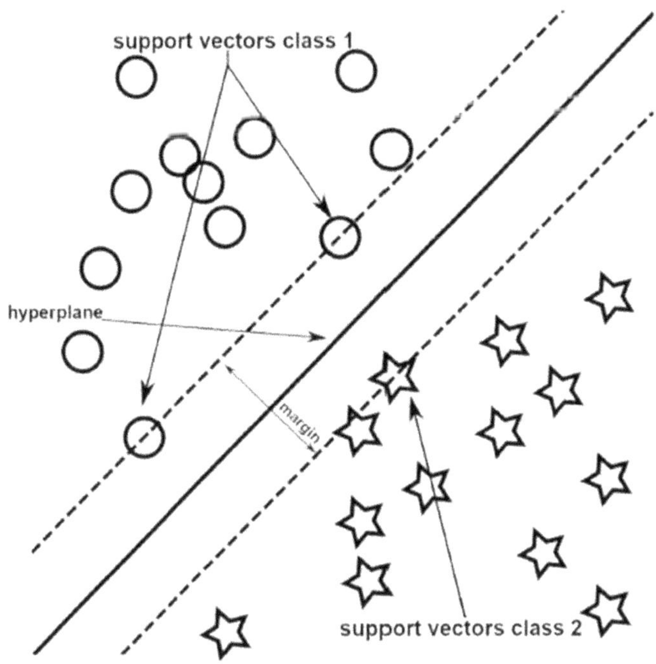

Figure 4.1 A simplified example of the support vectors and samples of two classes (adapted from [10]).

classification of objects [6–9]. As shown in Figure 4.1, this method involves finding the hyperplane, which separates objects from two different classes. In order to assign the objects to several classes, separate binary classification training is needed. The support vector machines are the nonprobabilistic binary classifier such that for a given training set (xi,yi) $y\epsilon\{\neq 1,1\}$. Such algorithms build a model that allocates new examples to one of the categories. Further, this model signifies a hyperplane with a classification boundary for the training data points. The novel data points are projected in the hyperplane such that for an optimal model, these data points extend on either side of the class boundary [10].

The hyperplane can be a plane, line, or point. It depends on the object's dimensions. Its objective is to state the class membership by evaluating which side of the boundary an object falls on. In some cases, an aim may have the properties of other classes, leading to the creation of a soft margin, where objects can be situated outside the well-defined cluster without distressing the position of the hyperplane. Further, if a single hyperplane cannot separate two data clusters, a kernel function may be employed to transform the data and obtain a linear separation of the classes [11].

4.2.5 Adaboost

Adaboost is a machine learning algorithm that is being used for binary classification and regression jobs. Adaboost is a fairly new non-linear machine learning algorithm [12]. It is a boosting technique that iteratively trains multiple weak models on different subsets of the training data and allocates higher weights to the misclassified instances in each iteration. This is illustrated in Figure 4.2. The algorithm focuses more on the misclassified samples in the following iterations, permitting the weak models to learn from their errors and develop on their enactment. Finally, they are pooled to form a single strong classifier [13]. This method is widely used in computer-aided diagnosis (CAD). Adaboost can also support medical practitioners to make critical decisions regarding their patients' health and disease conditions, such as Alzheimer's disease, diabetes, hypertension, and various cancers [14]. Moreover, adaboost can provide assistance to improve the correctness in

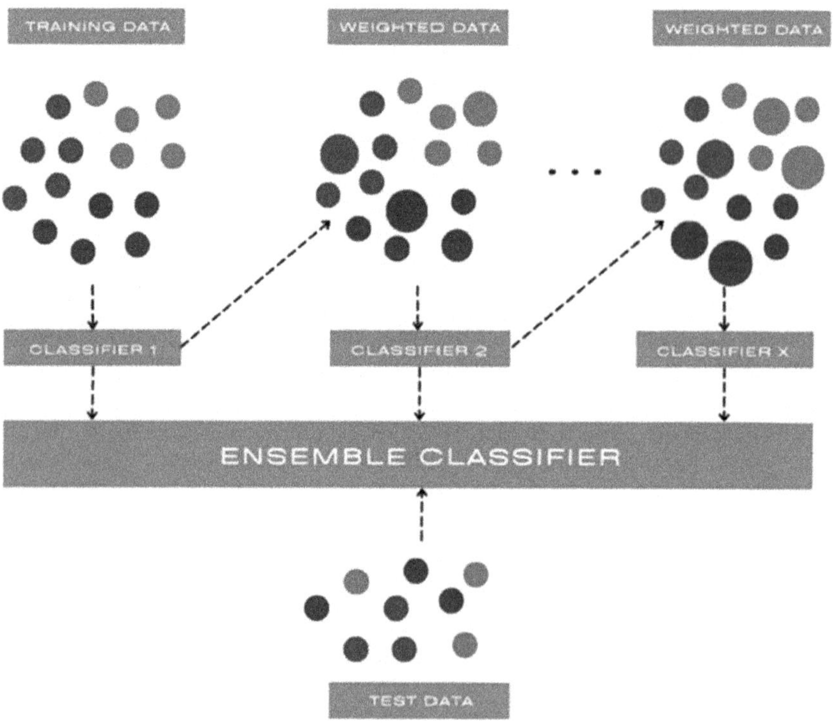

Figure 4.2 The illustration of the adaboost algorithm: Different sizes of the circles stand for samples with more associated weights; various colors of the circles stand for different subsets of data (adapted from [10]).

classifying diseases, patient outcomes, and predicting and detecting abnormalities in medical images [15–16].

4.2.6 T-distributed stochastic neighbor embedding

T-distributed stochastic neighbor embedding is a newly developed non-linear method [17]. The t-SNE algorithm is mostly employed for the visualization of a dataset in the low-dimensional latent space by reducing the dimensionality. The decline in the dimensionality can be realized by creating a probability distribution over the pairs of higher-dimensional data points, which is followed by building a similar probability distribution for the points in lower-dimension and minimizing the Kullback-Leibler divergence between the two distributions. The remarkable ability of this method is to decrease the dimensionality of the data while preserving the metric properties' consistency. The t-SNE method has initiated numerous applications in the field of music analysis, biomedical research, and computer security research, as well as obtaining the latent space representations learned by the ANN [18–21].

4.2.7 Artificial neural networks

AI has gained significant attention over the past decade, with machine learning and deep learning [22, 23]. ANNs look like the human brain and encompass multiple "neurons" or perceptrons that process and transmit information, as illustrated in Figure 4.3. The ANN consists of interconnected neurons. The existing technological age grants the onset of massive datasets, which establish the necessity of faster, automated, robust, and reliable methods to analyze them. The solution to the previous statement is to

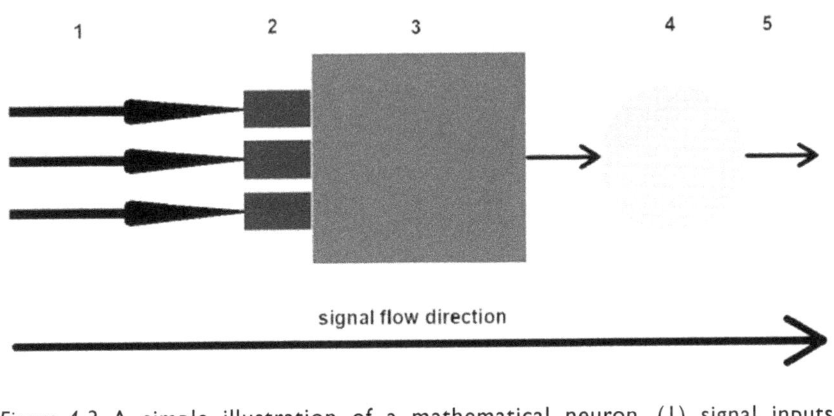

Figure 4.3 A simple illustration of a mathematical neuron. (1) signal inputs, (2) scales, (3) adder, (4) activator (activation function), and (5) signal output (adapted from [10]).

employ the latest computing methodology of ANNs, which are stimulated by the biological neural networks constituting animal brains. The ANNs are based on the assembly of fundamental units called artificial neurons. These neurons are proficient in both transmitting and receiving signals through their connections, as is observed in their biological equivalents. Starting with inputting data, such as text, sound, or images, we can understand the functioning of neurons, as shown in Figure 4.4. This data navigates the network processed by means of successive layers of neurons until reaching the output. Each layer comprises multiple neurons that process the input data. These connections form the parameters of the network. The receiving neuron processes the signal and communicates with the next neuron in the form of real numbers. The synaptic links, which are the weights and bias parameters, signify the memory of ANN. These parameters vary endlessly with the learning procedure. Usually, the neurons are systematized into layers such that different layers execute diverse alterations on their input. These signals enter the ANN from the input layer and travel through with consecutive processing to the output layer. To understand the functionality, see Figures 4.3 and 4.4.

Neural networks are widely used in speed recognition, natural language procession, image recognition, and stock price prediction. We can distinguish between the ANNs and deep neural networks (DNNs) based on the number of neural networks. ANNs are neural networks consisting of a

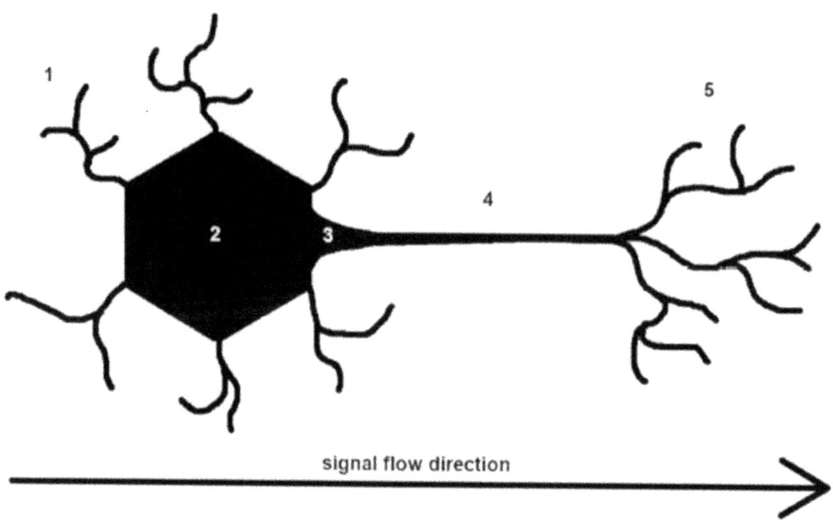

Figure 4.4 A simple illustration of a human neuron. (1) dendrites, signal input site; (2) nucleus of the neutron; (3) zone of initiation (where the action potential of the neutron is formed); (4) axon; and (5) axon terminals (which form connections with other cells and are the sites of signal output; adapted from [10]).

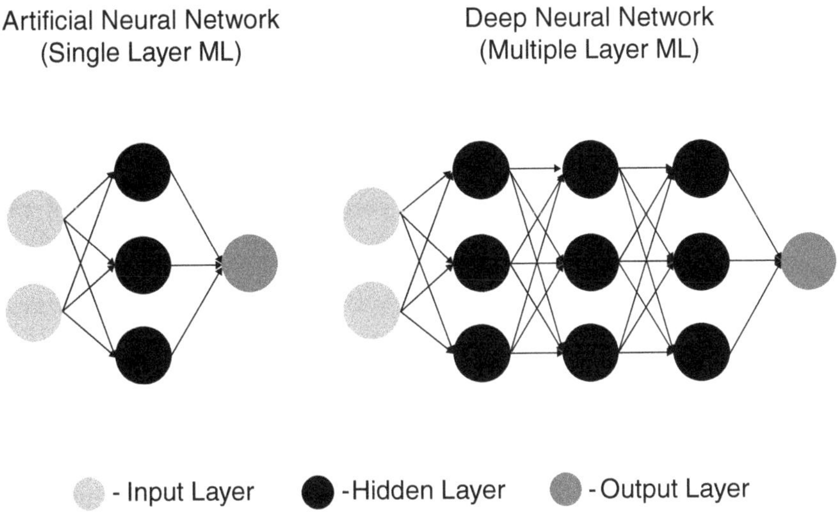

Figure 4.5 The graphical representation of an ANN and a DNN (adapted from [10]).

single hidden layer, and DNNs are neural networks that have multiple hidden layers. The schematic representation of differences between ANNs and DNNs is shown in Figure 4.5. This permits the network to mimic and recognize more complex and intangible behaviors.

4.3 AI IN BIOPHOTONICS

AI-based techniques have been spreading in most scientific areas and have become an exciting focus in photonics, modeling and simulation research, and beyond in recent years [24–26]. The potential applications of AI methods exclusively in integrated quantum photonics were presented earlier [27]. In the photonics field, device modeling can be broadly separated into forward and inverse modeling. The forward modeling and inverse design by AI can realize high efficiency and better accuracy for the photonics components. By employing the AI-assisted electronic circuit design for the photonics components, more innovative photonics applications have materialized.

In a developed immune system, T and B lymphocytes are present that play crucial roles in protecting the host from the infection. It is significant and technically challenging to classify cell types and their activation status in living and intact immune cells deprived of staining or killing the cells. This can be achieved using Raman spectroscopy. Discriminating between living

T cells and B cells, and visualizing the activation status of living T cells without labeling has been reported already [28]. Further, AI is proceeding into biomedical science in various ways, from cultivating image-based diagnostics to the engineering approaches for refining movement related to birth defects, injury, and neurological or cardiovascular diseases. Recently, development in several experimental methodologies to understand the high complication of the human genome and proteome has steered toward the generation of growing digital information, henceforth bioinformatics, which connects computer science, chemistry, and biology. The evolving technology of AI, including machine learning and ANNs, is at the core of biomedical research.

Furthermore, the role of AI-based technologies in various biological research fields has been reported [29]. The connection of AI-based tools to the various levels of modern biomedicine is illustrated in Figure 4.6. The enormous amount of biological data has established machine learning as the prospective instrument for the various fields, including transcriptomics, structural biology, system biology, and genomics [30]. Several computational genomics tools have been established to identify a given DNA

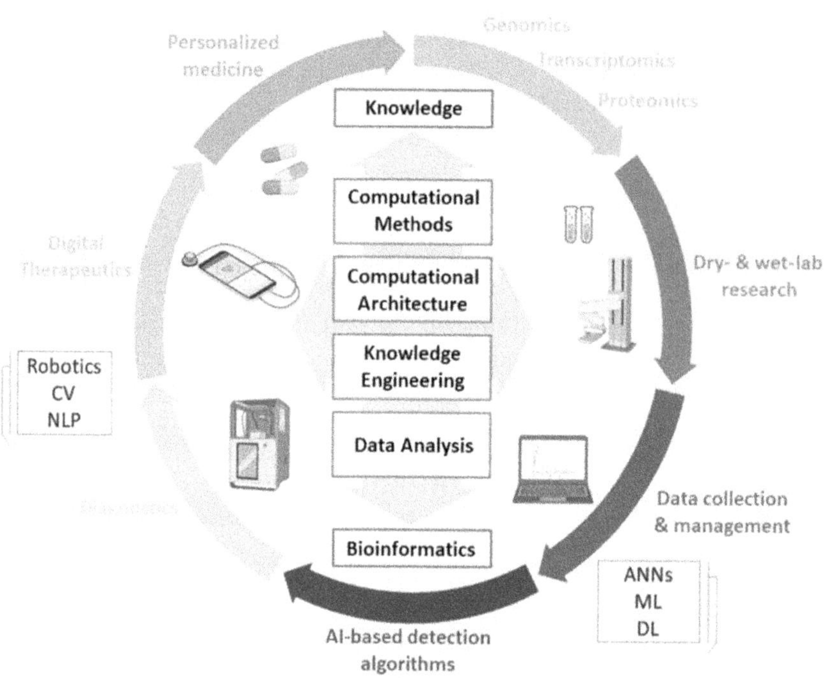

Figure 4.6 The illustration depicting the connection of AI-based tools in the various levels of the modern biomedicine (adapted from [29]).

sequence, envisage genes, define their functions, and valuate the location of protein-coding genes [31].

Machine learning and ANNs have gained substantial ground by providing thorough insights into the analysis of the proteome and interactome [32]. Computational machine learning algorithms, including support vector machines and ANNs, can be classified into two different groups: (1) methods that use the sequence-based approach and (2) methods that employ the structure-based approach [33, 34]. In the study of proteomics, machine learning software can be used for the understanding of proteomic mass spectrometry data, the detection of discrete proteins, and the inspection of their prospective, such as cancer biomarkers [35, 36]. For time-of-flight mass spectrometry, a sequential–based machine learning tool called "ML-ToF" has been established for identifying mass spectrometry peaks and analyzing datasets to match the produced labels to the established ones, thus recognizing amino acid sequences [37]. The structure-based machine learning algorithms for interactome studies have been established for prediction of the hotspot residues contributing to protein-protein interactions [38, 39]. Moreover, the Interactome INtegrated Structural Interactome and genomic data browser (INSIDER) is applicable for determining protein-protein interplays, including enriched information for mutations available in human diseases, such as cancer [40]. INSIDER is an additional structure-based machine learning tool. Machine learning is used to build models that envisage the binding affinity of every regulatory element, DNA, RNA, protein molecule, and region of interest.

Over the past decade, machine learning and deep learning techniques for molecule design have progressed and been established to address target-ligand interactions and recognize potential drugs and estimate their biological activity. Machine learning is applied in all chemical structures of the potential drug and in investigating the drug's effectiveness. The five phases of drug design and development processes are illustrated in Figure 4.7.

Various applications of AI techniques include deep learning models like the autoencoders, convolutional neural network (1D and 2D), and generative adversarial networks for analyzing biophotonic data. These applications have motivated the exploration of all variations of the deep learning models for data analysis purposes. The first issue is the potential application of deep learning models for Raman spectroscopic data, such as autoencoders or deep convolutional neural networks that have been used for spectra classification and image segmentation. Nevertheless, the deep convolutional neural networks can also be employed in the analysis of Raman spectroscopic data, including baseline correction and despiking for calibration. Similarly, one of the advanced DNNs, like the "recurrent neural network" (RNN), can also be examined for the preprocessing of Raman spectra.

The Fourth Industrial Revolution has headed toward the application and development in the field of health care. Monitoring sensors can be characterized by the intelligence and digitalization. Such sensors for health care

Figure 4.7 Phase of drug design and development process. The phases are divided into the basic research and the early discovery phase (step 1), pre-clinical trials (step 2), the clinical development phase (step 3), the Food and Drug Administration review process (step 4), and, finally, the post-marketing surveillance phase (step 5; adapted from [29]).

have wide applications in the medical care, sports, elderly care, personal health management, and other fields, providing people with more appropriate and real-time health amenities. The advancement of AI has provided powerful tools and algorithms for data processing and analysis, enabling intelligent health monitoring and achieving high-precision predictions and decisions. The fusion of AI and artificial sensors will deliver added intelligent, secure, and convenient services for next-generation health care and biomedical applications [41].

The general idea of the development of artificial sensors empowered by AI is illustrated in Figure 4.8. The left side of the figure exemplifies the road map of the sensing technologies. The AI technologies are illustrated in the right side of the figure. This figure further illustrates the development of the expanded applications in self-powered, implantable, wearable sensors and smart home, machine learning, volatile organic compounds (VOC) sensing, and human-machine interfaces. Various types of advanced AI-enhanced

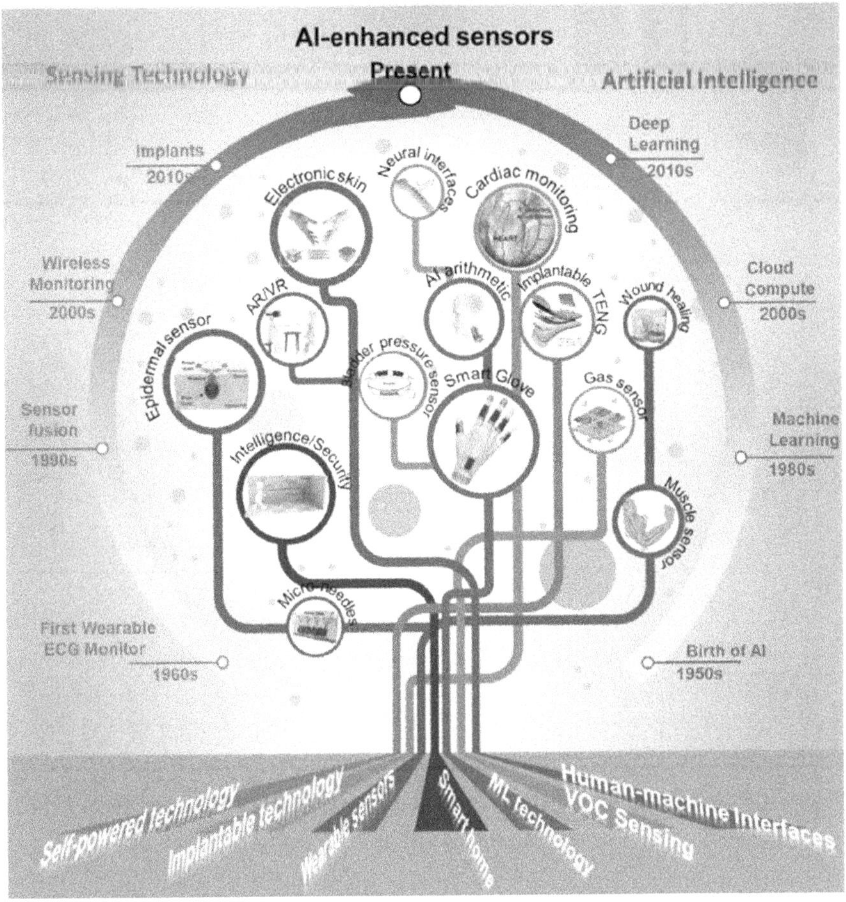

Figure 4.8 Overview of the development of artificial sensors enabled by intelligent technologies (adapted from [41]).

wearable glove sensors are illustrated in Figure 4.9. The scalable and affordable tactile glove (STAG) capable of weight estimation, object identification, and hand pose recognition is shown. The STAG encompasses a sensing sleeve having 584 piezoresistive sensors, a knit glove, and readout electronics that utilize the ResNet-18 architecture to train the model and eventually identify 26 different objects. Such techniques demonstrate that a meaningfully larger amount of the information becomes obtainable for examining interactions at a more thoughtful level, thanks to the developments in the flexible electronic components and the support of deep learning methods, enabling the future design and development of next-generation flexible and wearable electronics and the systems. A flexible quadruple tangible sensor permits a robotic hand to recognize grasped objects of

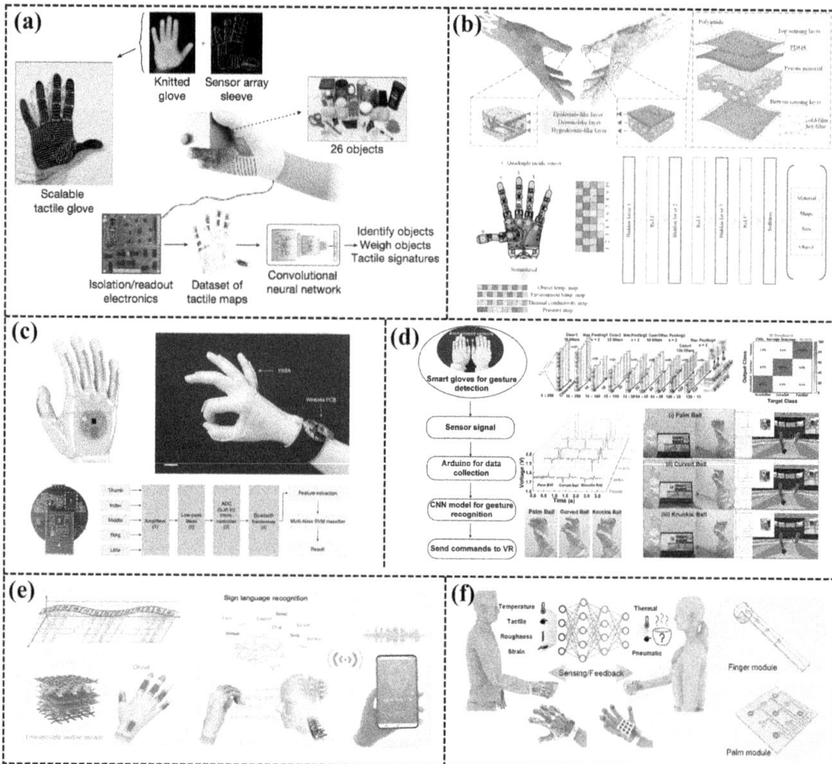

Figure 4.9 Advanced AI-enhanced wearable glove sensors (adapted from [41]).

changeable materials and shapes and further employs a multilayer perceptron (MLP) with three hidden layers to achieve automatic waste classification. The tactile sensors are able to notice multiple stimuli concurrently with minor cross-coupling errors, posing more perceptive features associated with the objects and improving recognition precision during the machine learning process. Henceforth, this proves their potential in markedly decreasing the load on individuals for environmental protection and sustainable progress in smart homes [41].

More interestingly, the piezoelectric and triboelectric sensors are able to generate self-produced voltage upon mechanical deformation. Such a feature of the sensors removes the requirement for external power sources. The triboelectric nanogenerator (TENG) strain sensors are based on a distinctive yarn structure [42]. The core of such a sensor unit includes a conductive yarn covering a rubber microfiber. Additionally, a self-powered, low-cost, and intuitive glove-based human-machine interaction (HMI) can be developed by conjoining superhydrophobic triboelectric textile sensors with machine learning [43–45]. This groundbreaking design permits complex gesture recognition and control in both virtual and real spaces while reducing the negative effects of humidity and sweat on performance. The communication system and sign language recognition based on TENG gloves' sensors have been reported [46]. The deep learning algorithm primarily recognizes word components, followed by the rebuilding of the original sentences, attaining respective accuracies of 82.81% and 85.58%. Unambiguously, the recognized word units can be restructured into a different sequence to frame fresh sentences. Simultaneously, the deep learning model classifies all the fundamental word elements in a new sentence and recommends the suitable translation. Hence the sentences not involved in the training dataset can be recognized. Finally, the sentence recognition results are envisioned in a virtual space where sign language customers can communicate by means of their familiar language, while nonsigners can directly input via their controlled virtual reality interface.

This innovation in identifying both existing and new sentences improves the practicality of the sign language recognition systems, marking a noteworthy step toward minimizing communication obstacles between sign language users and nonsigners.

The advancements and potential of AI-enhanced sensors in empowering next-generation health-care and biomedical platforms are shown in Figure 4.10. During the past few decades, there has been noteworthy progress in the design and development of the flexible sensors, which have established outstanding capabilities in reliably monitoring vital biomedical and physiological data, as well as daily activities. Contemporarily, the appearance of the self-powered sensors and energy harvesters has presented capable alternatives and supplementary methods in establishing a bodyNET framework [47].

The integration of AI into the sensor technologies holds immense promise for the future of health care and biomedicine. One of the key benefits of AI-enhanced sensor systems is the realization of personalized health care. By constantly monitoring physiological parameters and without a glitch integrating them with the individual patient's medical history, AI algorithms are able to generate personalized recommendations for early intervention, disease prevention, and chronic disease management. Furthermore, AI-enabled sensors can enable timely intervention, real-time monitoring of vital signs, and enabling rapid detection of the abnormalities, thus potentially saving lives. Enabled by the advanced therapeutic devices, including drug delivery patches and the neural interfaces, closed-loop sensing therapy systems arise as an encouraging platform to deliver superior benefits to patients suffering from chronic diseases. Further, the future of AI-supported sensors in biomedicine and health care appears even more encouraging. The conjunction of AI technology with self-sustainable Intelligence of Things systems will empower interminable and universal monitoring of an individual patient's health status, both outside and inside

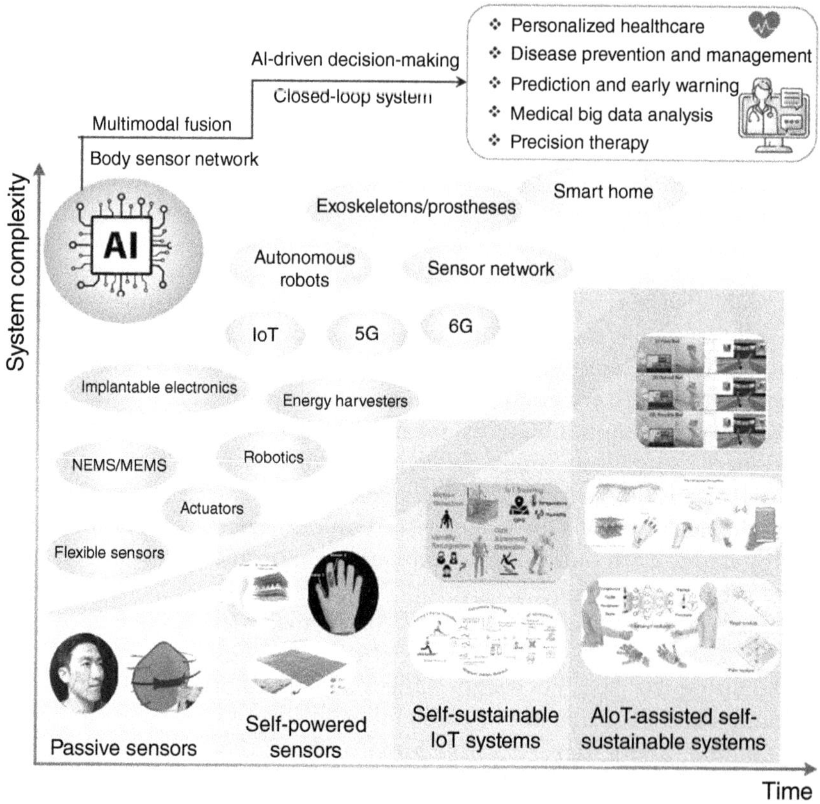

Figure 4.10 An overview of the technology development from discrete passive sensors to comprehensive and capable Artificial Intelligence of Things (AIoT)–assisted self-sustainable systems for diversified health-care and biomedicine applications (adapted from [41]).

of the body. The seamless integration of sensors, AI algorithms, and wireless communication technologies will empower patients to actively participate in their health-care management, and hence foster a proactive approach to well-being. Furthermore, the prevalent adoption of AI technologies in the medical research and clinical practices will stimulate the design and development of unconventional therapeutics and precision medicine tailored to individual patients. Nonetheless, the successful execution of AI-enhanced sensors in the biomedicine and health-care advances various challenges that should be judiciously addressed. Certifying data security and privacy, addressing regulatory and ethical considerations, and developing robust validation frameworks are crucial for the building trust in AI-based devices and systems. For example, AI-enhanced sensors process and collect massive amounts of individual health information, vigorous measures must be in place to protect patient privacy and prevent unsanctioned admittance or data breaches. By considering the role of AI-enhanced sensors in the creation of critical health-care decisions, compliance with regulatory frameworks becomes even more decisive.

REFERENCES

[1] Park, C., Took, C. C., Seong, J. K. Machine learning in biomedical engineering. *Biomed Eng Lett*, 8(1), 1–3 (2018).

[2] Ward, T.M., Mascagni, P., Madani, A., Padoy, N., Perretta, S., Hashimoto, D.A. Surgical data science and artificial intelligence for surgical education. *J Surg Oncol*, 124, 221–230 (2021).

[3] Mortani Barbosa, E.J., Gefter, W.B., Ghesu, F.C., Liu, S., Mailhe, B., Mansoor, A., Grbic, S., Vogt, S. Automated detection and quantification of COVID-19 airspace disease on chest radiographs: a novel approach achieving expert radiologist-level performance using a deep convolutional neural network trained on digital reconstructed radiographs from computed tomography-derived ground truth. *Investig Radiol*, 56, 471–479 (2021).

[4] Wang, Xishuo, Zhang, Qi, Xin, Xiangjun, Gao, Ran, Tian Qinghua, Tian, Feng, Wang, Chuxuan, Pan, Xiaolong, Wang, Yongjun, and Yang, Leijing. Robust weighted k-means clustering algorithm for a probabilistic-shaped 64qam coherent optical communication system. *Opt Express*, 27(26):37601–37613, 2019.

[5] Mari, Jean-Martial, Aung, Tin, Cheng, Ching-Yu, Strouthidis, Nicholas G., Girard, Michael J.A. A digital staining algorithm for optical coherence tomography images of the optic nerve head. *Transl Vis Sci Technol*, 6(1), 8 (2017).

[6] Boser, Bernhard E., Guyon, Isabelle M., Vapnik, Vladimir N. A training algorithm for optimal margin classifiers. In *Proceedings of the fifth annual workshop on Computational learning theory*, pages 144–152, 1992.

[7] Gunn, Steve R. et al. Support vector machines for classification and regression. *ISIS Technical Report*, 14(1), 5–16, 1998.

[8] Yang, Haiqin, Chan, Laiwan, King, Irwin. Support vector machine regression for volatile stock market prediction. In *International conference on intelligent data engineering and automated learning*, pages 391–396. Springer, 2002.

[9] Pal, Mahesh, Mather, P.M. Support vector machines for classification in remote sensing. *Int J Remote Sens*, 26(5):1007–1011 (2005).

[10] Kufel, Jakub, et. al., What is machine learning, artificial neural networks and deep learning? examples of practical applications in medicine. *Diagnostics*, 13, 2582 (2023).

[11] Noble, W.S. What is a support vector Mmachine? *Nat Biotechnol* 24, 1565–1567 (2006).

[12] Schapire, R.E., Freund, Y. *Boosting: Foundations and Algorithms*, *Adaptive Computation and Machine Learning Series*, MIT Press: Cambridge, MA, USA, 2012.

[13] Li, S., Zeng, Y., Chapman, W.C., Erfanzadeh, M., Nandy, S., Mutch, M., Zhu, Q. Adaptive boosting (AdaBoost)-based multiwavelength spatial frequency domain imaging and characterization for ex vivo human colorectal tissue assessment. *J Biophotonics*, 13, e201960241 (2020).

[14] Hatwell, J., Gaber, M.M., Atif Azad, R.M. Ada-WHIPS: explaining AdaBoost classification with applications in the health sciences. *BMC Med Inform Decis Mak*, 20, 250 (2020).

[15] Baniasadi, A., Rezaeirad, S., Zare, H., Ghassemi, M.M. Two-step imputation and AdaBoost-based classification for early prediction of sepsis on imbalanced clinical data. *Crit Care Med*, 49, e91–e97 (2021).

[16] Takemura, A., Shimizu, A., Hamamoto, K. Discrimination of breast tumors in ultrasonic images using an ensemble classifier based on the AdaBoost algorithm with feature selection. *IEEE Trans Med Imaging*, 29, 598–609 (2010).

[17] Laurens van der Maaten, Geoffrey Hinton. Visualizing data using t-SNE. *J Mach Learn Res*, 9, 2579–2605 (2008).

[18] Hamel, Philippe, Eck, Douglas. Learning features from music audio with deep belief networks. In *ISMIR*, volume 10, pages 339–344. Utrecht, The Netherlands, 2010.

[19] Jamieson, Andrew R., Giger, Maryellen L., Drukker, Karen, Li, Hui, Yuan, Yading, Bhooshan, Neha. Exploring nonlinear feature space dimension reduction and data representation in breast cadx with laplacian eigenmaps and-sne. *Med Phys*, 37(1), 339–351 (2010).

[20] Wallach, Izhar, Lilien, Ryan. The protein–small-molecule database, a non-redundant structural resource for the analysis of protein-ligand binding. *Bioinformatics*, 25(5), 615–620 (2009).

[21] Ali Eslami, S.M., Jimenez Rezende, Danilo, Besse, Frederic, Viola, Fabio, Morcos, Ari S., Garnelo, Marta, Ruderman, Avraham, Rusu, Andrei A., Danihelka, Ivo, Gregor, Karol, et al. Neural scene representation and rendering. *Science*, 360(6394), 1204–1210 (2018).

[22] Ward, T.M., Mascagni, P., Madani, A., Padoy, N., Perretta, S., Hashimoto, D.A. Surgical data science and artificial intelligence for surgical education. *J Surg Oncol*, 124, 221–230 (2021).

[23] Mortani Barbosa, E.J., Gefter, W.B., Ghesu, F.C., Liu, S., Mailhe, B., Mansoor, A., Grbic, S., Vogt, S. Automated detection and quantification of COVID-19 airspace disease on chest radiographs: a novel approach achieving expert radiologist-level performance using a deep convolutional neural network trained on digital reconstructed radiographs from computed tomography-derived ground truth. *Investig Radiol*, 56, 471–479 (2021).

[24] Alagappan, Gandhi, Rong Ong, Jun, Yang, Zaifeng, Long Ang, Thomas Yong, Zhao, Weijiang, Jiang, Yang, Zhang, Wenzu, Png, Ching Eng, Leveraging AI in photonics and beyond. *Photo-Dermatology* 9, 75 (2022).

[25] Jiang, J., Chen, M., Fan, J.A. Deep neural networks for the evaluation and design of photonic devices. *Nat Rev Mater* 6, 679–700 (2021).

[26] Ma, W., Liu, Z., Kudyshev, Z.A., Boltasseva, A., Cai, W., Liu, Y. Deep learning for the design of photonics structures. *Nat Photonics*, 15, 77–90 (2021).

[27] Pfau, D., Spencer, J.S., Matthews, A.G., Foulkes, W.M.C. Ab initio solution of the many-electron Schrödinger equation with deep neural networks. *Phys Rev Res*, 2, 033429 (2020).

[28] Ichimura, T. et al. Non-label immune cell state prediction using Raman spectroscopy. *Sci Rep* 6, 37562 (2016).

[29] Athanasopoulou, K., Daneva, G.N., Adamopoulos, P.G., Scorilas, A. Artificial intelligence: the milestone in modern biomedical research. *Biomedinformatics*, 2, 727–744 (2022).

[30] Larranaga, P., Calvo, B., Santana, R., Bielza, C., Galdiano, J., Inza, I., Lozano, J.A., Armananzas, R., Santafe, G., Perez, A., et al. Machine learning in bioinformatics. *Brief Bioinform*, 7, 86–112 (2006).
[31] Mathe, C., Sagot, M.F., Schiex, T., Rouze, P. Current methods of gene prediction, their strengths and weaknesses. *Nucleic Acids Res*, 30, 4103–4117 (2002).
[32] Swan, A.L., Mobasheri, A., Allaway, D., Liddell, S., Bacardit, J. Application of machine learning to proteomics data: Classification and biomarker identification in postgenomics biology. *OMICS*, 17, 595–610 (2013).
[33] Lise, S., Buchan, D., Pontil, M., Jones, D.T. Predictions of hot spot residues at protein-protein interfaces using support vector machines. *PLoS One*, 6, e16774 (2011).
[34] Preto, A.J., Matos-Filipe, P., de Almeida, J.G., Mourao, J., Moreira, I.S. Predicting hot spots using a deep neural network approach. *Methods Mol Biol*, 2190, 267–288 (2021).
[35] Lee, E.S., Durant, T.J.S. Supervised machine learning in the mass spectrometry laboratory: A tutorial. *J Mass Spectrom Adv Clin Lab*, 23, 1–6 (2022).
[36] Yates, J.R., Ruse, C.I., Nakorchevsky, A. Proteomics by mass spectrometry: approaches, advances, and applications. *Annu Rev Biomed Eng*, 11, 49–79 (2009).
[37] Wei, Y., Varanasi, R.S., Schwarz, T., Gomell, L., Zhao, H., Larson, D.J., Sun, B., Liu, G., Chen, H., Raabe, D., et al. Machinelearning-enhanced time-of-flight mass spectrometry analysis. *Patterns*, 2, 100192 (2021).
[38] Moreira, I.S., Koukos, P.I., Melo, R., Almeida, J.G., Preto, A.J., Schaarschmidt, J., Trellet, M., Gumus, Z.H., Costa, J., Bonvin, A. SpotOn: High accuracy identification of protein-protein interface hot-spots. *Sci Rep*, 7, 8007 (2017).
[39] Qiao, Y., Xiong, Y., Gao, H., Zhu, X., Chen, P. Protein-protein interface hot spots prediction based on a hybrid feature selection strategy. *BMC Bioinform*, 19, 14 (2018).
[40] Meyer, M.J., Beltran, J.F., Liang, S., Fragoza, R., Rumack, A., Liang, J., Wei, X., Yu, H. Interactome INSIDER: A structural interactome browser for genomic studies. *Nat Methods*, 15, 107–114 (2018).
[41] Wang, Chan, He, Tianyiyi, Zhou, Hong, Zhang, Zixuan, Lee, Chengkuo. Artifcial intelligence enhanced sensors – enabling technologies to next-generation healthcare and biomedical platform. *Bioelectronic Med*, 9, 17 (2023).
[42] Zhou, Z, Chen, K, Li, X, Zhang, S, Wu, Y, Zhou, Y, Meng, K, Sun, C, He, Q, Fan, W, Fan, E, Lin, Z, Tan, X, Deng, W, Yang, J, Chen, J. Sign-to-speech translation using machine-learning-assisted stretchable sensor arrays. *Nat Electron* 3, 571–578 (2020).
[43] Wen, F, He, T, Liu, H, Chen, H-Y, Zhang, T, Lee, C. Advances in chemical sensing technology for enabling the next-generation self-sustainable integrated wearable system in the IoT era. *Nano Energy*, 78, 105155 (2020).
[44] Wen, F, Sun, Z, He, T, Shi, Q, Zhu, M, Zhang, Z, Li, L, Zhang, T, Lee, C. Machine learning glove using self-powered conductive superhydrophobic triboelectric textile for gesture recognition in VR/AR applications. *Adv Sci (Weinh)*, 7, 2000261 (2020).
[45] Wen, F, Sun, Z, He, T, Shi, Q, Zhu, M, Zhang, Z, Li, L, Zhang, T, Lee, C. Machine learning glove using self-powered conductive superhydrophobic triboelectric textile for gesture recognition in VR/AR applications. *Adv Sci*, 7, 2000261 (2020).

[46] Wen, F, Zhang, Z, He, T, Lee, C. AI enabled sign language recognition and VR space bidirectional communication using triboelectric smart glove. *Nat Commun*, 12, 5378 (2021).
[47] He, T, Lee, C. Evolving flexible sensors, wearable and implantable technologies towards BodyNET for advanced healthcare and reinforced life quality. *IEEE O J Circuits Sys*, 2, 702–720 (2021).

Chapter 5

AI-enhanced quantum optical devices

Than Singh Saini

5.1 INTRODUCTION

Artificial intelligence (AI) technologies are attractive and growing universal and extensive in a broad range of perspectives from commercial applications to cutting-edge scientific research. In AI, one can simulate human intelligence processes via machines, especially computer systems. In a true sense, AI is a part of our modern life, as it is quite tough to identify an area where accomplishments in the field of AI would not be used. AI technologies enable machines to learn useful processes, including object recognition, speech recognition, natural language processing, machine vision, and digital personal assistants. In this context, quantum optics and quantum-based technologies have been found in prominent positions since they enable a speed up in learning processes [1–4]. It has been perceived that the demand for the algorithms that enable machines to learn quickly and efficiently have increased day-by-day as the field of AI has advanced. Quantum machine learning is a fast rising field connecting AI and quantum technology.

The areas of AI and quantum technologies have grown in parallel, and have established considerable potential to supplement each other. Incorporation of both fields refers to the use of AI techniques to advance algorithms for quantum physics and quantum computing, as well as the practice of quantum computing to enhance AI applications. Quantum computing has the potential to revolutionize various fields. Controlling quantum systems is scandalously difficult, which is one of the major obstacles standing in the way of extensive use of quantum computing. AI has unlocked new paths for automated control of the quantum systems. Specifically, the application of AI can provide invaluable insight into the complex and multifaceted domain of quantum physics to accelerate the discovery of quantum physics laws, and can potentially alleviate challenges that have been historically associated with quantum computing and quantum communication. Alternatively, quantum computing can also be used to enhance AI applications. For example, quantum computing can be used to accelerate the training of neural networks, which are being used in machine learning. Concurrently, a series of advancements in quantum technology can assist in

DOI: 10.1201/9781003591030-5

driving innovation in the realm of machine learning by empowering the improvement of innovative hardware, frameworks, and algorithms.

The major points of work in the field of AI for quantum includes issues about data security, data privacy, and requirement for responsible development, as well as deployment of such technologies. AI for quantum technology works synergistically and transforms the fields of machine learning and pattern recognition. Also, the limitations of the classical computing in accommodating the current computational process of complex AI models lead to the requirement of quantum computing. Moreover, the effectiveness of quantum computing to report complex optimization difficulties and enhance data analysis leads to prospective benefits in AI and related fields. Furthermore, cross-disciplinary collaboration and reciprocal enhancement between AI and quantum technology is necessary to fully leverage their potential synergies and promote mutual evolution.

For the development of quantum algorithms, some AI techniques have been used, including variational quantum algorithms and quantum generative adversarial networks. Variational quantum algorithms are a category of algorithms that employ AI techniques for optimization of the quantum circuits. These sorts of algorithms can learn to adjust the parameters of a quantum circuit in real time, sanctioning the development of efficient quantum algorithms. However, quantum generative adversarial networks are a sort of generative model that uses quantum circuits to produce data. Several researchers have used quantum generative adversarial networks to develop quantum algorithms that can create samples from complex quantum systems, such as molecules. Such types of AI techniques have revealed promising results in the development of quantum algorithms, but there is still demand for further research in this area. This chapter provides an inclusive overview of the mutual relationship between AI and quantum technologies, methodologies that can provide progress in the development of quantum technologies, and the power of AI.

5.2 HOW DOES AI WORK?

AI requires a foundation of dedicated software and hardware for writing and training machine learning algorithms. Various programming languages, such as Python, R, C++, Julia, and Java, have features popular with AI developers. In a broad spectrum, AI systems work by feeding large amounts of labeled training data and analyzing data for patterns and correlations. The system use the patterns to make conjectures about imminent states. In this manner, a chatbot that is served examples of text can learn to produce authentic exchanges with people, or the tool identifying images can learn to recognize and describe objects in images by running through millions of examples. Moreover, novel generative AI techniques are able to generate realistic text, images, media, and music.

5.3 MACHINE LEARNING

Machine learning is a subset of AI that is primarily concerned with the advancement of algorithms that allow a computer to understand or learn from the data and past experiences on their own. In 1959, the term "machine learning" was first familiarized by *Arthur Samuel*. Machine learning can be concisely defined as follows:

> Machine learning enables a machine to automatically learn from data, improve performance from experiences, and predict things without being explicitly programmed.

With the help of training data (i.e., sample historical data), machine learning algorithms construct mathematical models that aid in creating predictions or judgments without being overtly programmed. Machine learning brings statistics and computer science together for generating predictive models. Machine learning builds or uses the algorithms that learn from the training data. The more information we make available, the higher the performance. In other words, we can say that a machine has the skill to learn if it can progress its performance by gaining more data.

Machine learning can be classified into three main parts: (*i*) supervised learning, (*ii*) unsupervised learning, and (*iii*) reinforcement learning. The pictorial representation of the classification of machine learning is illustrated in Figure 5.1.

5.3.1 Supervised learning

Supervised learning is a method of machine learning. In this method, the models are trained using labeled data, and on the basis of that data, machines envisage the output. The labeled data means some input data is already tagged with the correct output. Supervised learning can be used for

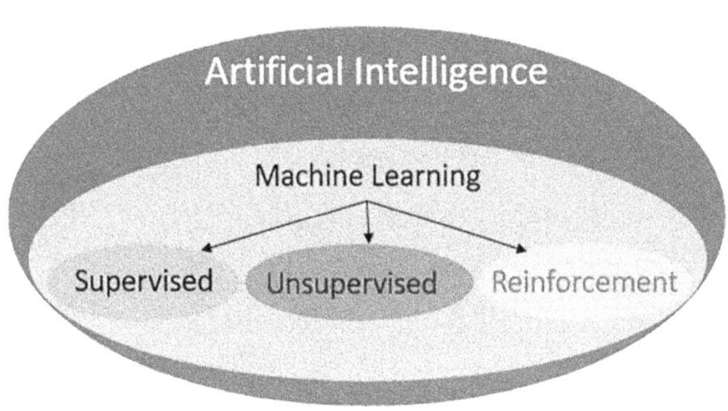

Figure 5.1 The illustration of classification of machine learning.

classifying data. In this learning, models are required to find the mapping function to map the input variable (X) with the output variable (Y), i.e., Y = f(X). In the similar way as a student in class learns a subject in the presence of a teacher, supervised learning requires supervision to train the model.

For example, as shown in Figure 5.2, we have a collection of different images. The task of the supervised learning model is to identify the animals from various images and discriminate dog or cat images. Accordingly, to recognize the image in supervised learning, we have to provide the input data, as well as the output. It means we can train the model by the characteristic like color, shape, and size of each animal. As soon as the training is finalized, we will trial the model by providing a new set of animal. Then the model will recognize the animal and envisage the output by means of a suitable algorithm. In the real-life experience, supervised learning can be employed for identification and classification of images, risk assessment, spam filtering, fraud detection, etc. The supervised learning can be grouped further in two categories of algorithms: (*i*) classification and (*ii*) regression.

5.3.2 Unsupervised learning

As in supervised machine learning, models are trained using labeled datasets. However, there may be some cases in which we do not have labeled data and are required to discover the hidden patterns from the available dataset. Hence, to solve such types of problems in machine learning, we require unsupervised learning procedures. In this type of machine learning, models are trained using unlabeled datasets and are permitted to act on that data without supervision. Unsupervised learning can be used for anomaly detection. As illustrated in Figure 5.2, an unsupervised learning algorithm performs the task by clustering the image dataset into the groups based on similarities between images. As an alternative, models themselves notice the hidden patterns and acumens from the available data. This can be compared to learning which takes place in the human brain while learning new things. There are two types of unsupervised learning: (*i*) clustering and (*ii*) association.

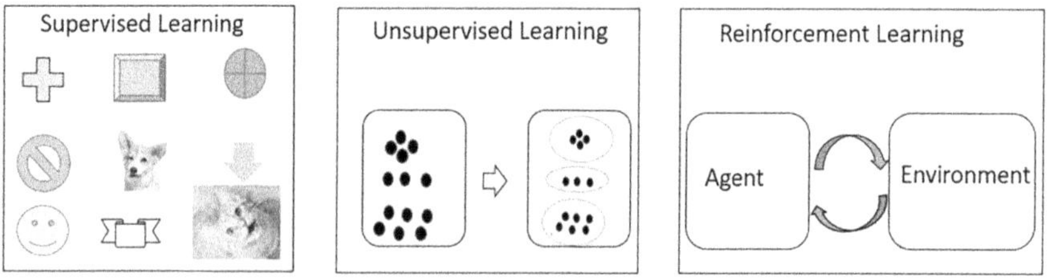

Figure 5.2 The illustration of supervised, unsupervised, and reinforcement learning processes.

Clustering is the technique of grouping the objects into clusters so that the objects with the most similarities remain in a group and have less or no similarities with the objects of a different group. Cluster analysis discovers the commonalities among the data objects and classifies them according to the occurrence and nonexistence of those commonalities. Association is the unsupervised learning method through which we can find out the relationships among various variables in the large database. Association regulates the set of items that occur together in the dataset. This technique can make the marketing approach more operational. A classic example of association technique is the "market basket analysis."

5.3.3 Reinforcement learning

Reinforcement learning is based on the exchange of feedback between an agent and the environment. Actually, it does not rely on data, while, supervised learning and unsupervised learning do. In this method, a learning agent gets an incentive for each correct action and gets a penalty for each wrong action. The agent robotically studies this feedback and progresses its performance. In reinforcement learning, the agent intermingles with the environment and explores the environment. The objective of an agent is to achieve more reward points, and in this way, it advances its performance. An example of reinforcement learning is the robotic dog in Figure 5.2, which spontaneously understands the movement of its arms.

5.4 QUANTUM MACHINE LEARNING

It has been found that quantum mechanics is the most powerful theory for modeling an extensive range of physical phenomena from subatomic to astral scales with precise accurateness. The spooky action of entanglement in the Einstein-Podolsky-Rosen pair [5], has delivered the resources for quantum information processing tasks. Quantum machine learning (QML) is a promptly developing new field of research at the border of AI, quantum technology, and quantum information science. Some milestones achieved up-to-now in the field of QML have been summarized in Table 5.1.

With the help of machine learning, we can model various physical systems, including quantum, gravitational, and statistical, using artificial neural networks, which may lead to the advancement of a new framework for fundamental physics. Machine learning supports solving problems of fundamental, as well as applied, quantum physics [27]. Furthermore, it can prove the quantum behavior of some experimental correlations with the device-independent test of Bell nonlocality, even without an accurate depiction of a physical apparatus and merely based on measured data [28]. Specifically, employing generative algorithms to blend automatically numerous multilayer perceptrons (a machine learning approach) can

Table 5.1 Milestones of the QML achievements

Year	*Milestones*	*Reference*
2007	Quantum clustering algorithm	[6]
2009	Quantum algorithm for linear systems of equations	[7]
2014	Quantum principal component analysis	[8]
2014	Quantum support vector machine	[9]
2014	Quantum approximate optimization algorithm	[10]
2014	Variational quantum eigensolver on a photonic quantum processor	[11]
2016	Machine learning creating quantum experiments	[12, 13]
2018	Quantum convolutional neural networks	[14, 15]
2018	Quantum error correction with machine learning	[16, 17]
2019	Graph neural networks detecting quantum advantage	[18, 19]
2021	Quantum reinforcement learning	[20, 21]
2022	Hybrid quantum computing-classical machine learning	[22]
2022	Practical industry applications: hybrid quantum neural networks for car classification, drug response prediction, and cancer target identification	[23–25]
2023	Maximizing supercontinuum bandwidths in gas-filled, hollow-core fibers using artificial neural networks	[26]

detect and quantify the nonlocality, as well as its quantum (or postquantum) nature [29–30].

5.5 AI FOR QUANTUM TECHNOLOGY

Nowadays, there is an enormous demand to integrate AI and quantum technology since it would be a critical region of exploration with enormous potential in future. The application of AI to quantum technology provides a capable means to address complex quantum problems. Specifically, if we think about it, Moore's Law approaches its physical limits. Further, quantum technology may serve to accelerate the advancement of AI at the application end of the industry. The relationship of quantum technology and AI is dignified to become a stimulating driver of economic and social development, representing various noteworthy opportunities for the future growth and innovations [31].

The superiority of AI in solving various challenges related to quantum physics has been well-known and documented in various reports. It is expected that the range of applications will increase order in the future. Some of them include quantum phase classification, error correction, and quantum communication. Likewise, AI and quantum computing have the potential to impact several other areas beyond machine learning and the optimization. Some potential applications of AI and quantum computing in the current age are given next:

5.5.1 AI in meta-optics

The synergy of AI and meta-optics has significantly promoted both fields. In the field of meta-optics, we try to advance flat optics with innovative functions and light-manipulation capacities [32]. Meta-optics is an optical technology that has emerged in recent years. Metasurfaces are advanced flat optical devices that can manipulate electromagnetic waves or light. The metasurfaces consist of a particularly arranged array of meta-atoms. The artificial meta-atoms are designed and computed to control the phase, amplitude, and polarization of the incident electromagnetic waves. The new wavefront can be reconstructed by metasurfaces. The development of AI and meta-optics is growing exponentially. The growth of AI and meta-optics is illustrated in Figure 5.3. Also, the illustration of the fabrication methods for metasurfaces is given in Figure 5.4.

The technology for the fabrication of metasurfaces has been well-developed and selected for various purposes, including subwavelength scale,

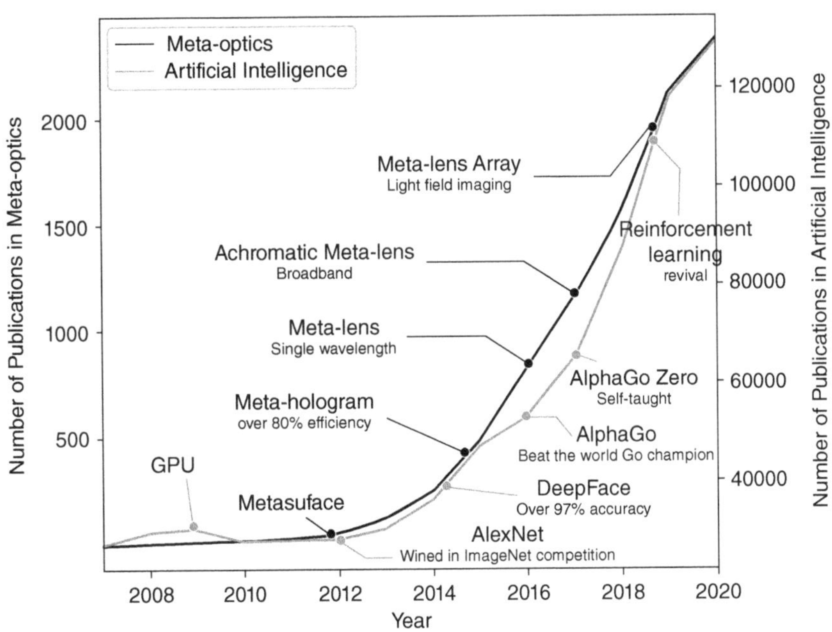

Figure 5.3 Diagram showing the development trend of AI and meta-optics. The number of publications per year was collected from a Web of Science search (adapted from [32]).

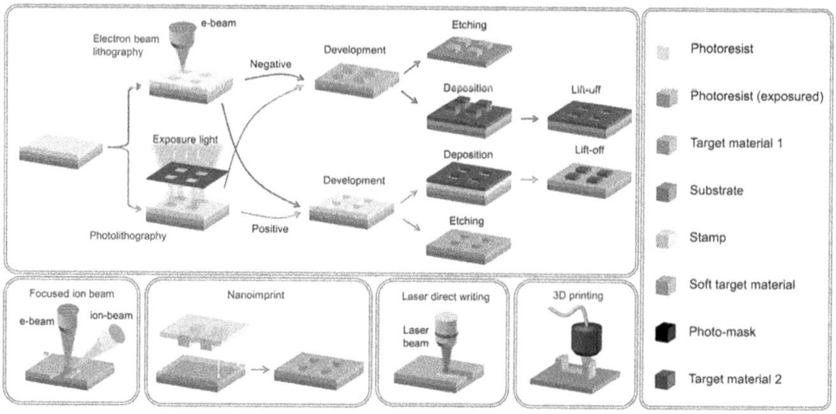

Figure 5.4 Illustration of the fabrication methods for metasurfaces (adapted from [32]).

large area, structure carving, high aspect ratio, high throughput, etc. The utmost commonly used fabrication methods are photolithography, laser direct writing, focused ion beam (FIB) lithography, electron beam lithography (EBL), nanoimprinting, and 3D printing. In the semiconductor manufacturing industry, photolithography has been an important fabrication technology for mass manufacturing. Photolithography employed a photomask and an exposure light to transfer the pattern to the photoresistor [33]. The solubility of the exposed photoresistor layer changes to enable pattern generation after the developing process. The lift-off and etching processes can further sculpt the feature of the target material using the patterned photoresistor layer.

EBL is a maskless lithography process. A focused electron beam is utilized to directly write the pattern on the photoresistor layer. EBL technology is broadly used for meta-optics research due to its quick verification and customizable design. The medium used is a photoresist whether a photomask or electron beam is used to transfer the pattern. Combined with the choice of a positive or negative photoresist, etching and lift-off processes can transform positive and negative structure pattern generation. A high aspect-ratio nanostructure can be fabricated with the hard mask etching process.

In the FIB technique, an ion beam is generated by the use of an ion source and then focused on the surface of the sample. FIB lithography is a straightforward fabrication method in which the patterning and etching processes happen simultaneously [34]. The accelerated ion beam strips the surface atoms of the sample. FIB with a scanning electron microscope becomes a dual-beam system, which can realize visualization during the ion-beam writing. Nanoimprinting is mechanical lithography method that uses a mold to imprint onto a UV- or thermal-curable resist.

The nano-or microstructures can be formed through mechanical deformation. The mold can be employed recurrently to have the same pattern with a low cost and for getting a high throughput. Laser direct writing employed a focused beam of laser to make localized damage on the surface of materials. In the laser writing method, the pattern can be written directly, similar to the FIB method. The damage caused by the laser beam is controlled by tuning the laser power. Recently developed 3D printing technology is also a rapid prototyping manufacturing method of fabrication of meta-optical devices [35]. This type of method is known as the additive manufacturing method.

Meta-devices with innovative and superior optical functionalities have been generated to meet optical demand. The great advantages of meta-devices are their lighter weight, novel properties, compact size, high efficiency, broadband operation, better performance, data volume reduction, lower energy consumption, and complementary metal-oxide-semiconductor (CMOS) compatibility for mass production. Optical meta-devices have been well-developed for beam shaping [35, 36, 37], anomalous deflection and reflection [38, 39], polarization control and analysis [40, 41], holographic images [42, 43], non-linear effects [44–49], lasing [50, 51], tenability [52–54], color display [55], light focusing [56], imaging [57], image processing [58], beam shaping [59], multiplex color routing [60], sensing [61], encryption [62], spectrometry [63], optical characterization [64], communication [65], augmented reality [66], and quantum technology [67].

5.5.2 AI in the health-care industry

AI plays an important role in the health-care industry [68]. With the rapid development of AI for analysis of imaging, it seems that most of the radiology and pathology images will be examined at some point by the machine. Text and speech recognition are being employed for tasks such as patient communication and capture of clinical notes. In the health-care industry, the utmost common application is precision medicine. Based on several patient aspects and the treatment context, precision medicine doctors predict treatment protocol, which is likely suitable for a patient. The majority of machine learning and precision medicine applications necessitate a training dataset for which the outcome variable is known; this is known as supervised learning. To improve their precision and efficiency, AI would be incorporated into surgical robots. Surgical robots provide "superpowers" to the human surgeons, hence improving their capability to see and provide very precise and nominally invasive cuts and stitch wounds. Of course, important decisions are still made by the human surgeons. Most common surgical procedures in which robotic surgery is used includes prostate surgery, gynecologic surgery, head surgery, and neck surgery.

5.5.3 AI in chemistry and physics

The features of AI can be helpful in analyzing large amounts of chemical and physical data. Quantum computing can be very applicable to simulating complicated atomic structures. From the point of view of chemistry, possibly the most significant approach is the modeling of chemical reactions and the computer aided rational design of molecules and synthetic approaches. In this technique, one considers retro-synthetic pathways ascending from a given product until a set of precursors that are commercially accessible or synthetically known in the literature can be achieved. Such pathways are recorded on efficacy based on the number of reactions involved, intermediates, and reaction conditions. Machine learning and its close cousin, deep learning, have accompanied extraordinary advances in all the areas of physical sciences, especially chemistry, and not only classical variants of machine learning: even those trainable on near-term quantum hardware have been developed with promising outcomes. Such algorithms have revolutionized material design and performance of photovoltaics, electronic structure calculations of ground and excited states of correlated matter, computation of force fields and potential energy surfaces informing chemical reaction dynamics, reactivity inspired rational strategies of drug designing, and even classification of phases of matter with accurate identification of emergent criticality [69].

5.5.4 AI in radiology and drug discovery

Quantum computing can also be very useful to simulate complicated molecular structures and characteristics to accelerate drug discovery. Also, the features of AI can be helpful in analyzing large amounts of chemical and biological data to aid in the discovery process. It has been predicted that the health-care professions most likely to be automated are those that deal with digital information, pathology, and radiology. Technologies such as deep learning are creating the capability of diagnosing and categorizing images. In a radiology exercise, trained human physicians visually assessed medical images for characterization, detection, and monitoring of various diseases. AI methods excel at robotically identifying complex patterns in the imaging data and offering quantitative, rather than qualitative, valuations of radiographic features [70]. To speed up the effective treatment development at a low cost and reduce animal experiments, drug discovery is now adapting to novel technologies, such as informatics, data science, and AI. It has been noticed that AI is transforming drug discovery, industrial and academic scientists, and legislators. AI technologies such as machine learning, multi-property optimization, and generative chemistry have empowered various compounds to enter clinical trials. The use of AI in three pillars of drug discovery, including targets, diseases, and therapeutic modalities, has been discussed in a recently published article by Hasselgren et. al. [71].

5.5.5 Finance and economics

Quantum computing can improve financial modeling, portfolio management, and risk assessment [72]. AI is able to assist in the prediction of stock prices, as well as current and future market trends. AI and machine learning are useful for groups of finance scholarship, including (*i*) portfolio construction, valuation, and investor behavior; (*ii*) financial fraud and distress; and (*iii*) sentiment inference, forecasting, and planning.

5.5.6 Cybersecurity

Nowadays, security teams face many challenges, including an explosion of data, sophisticated cyberattacks, an expanding attack surface, and growing infrastructure complexity that hamper their capability to protect data, manage user access, and quickly detect and respond to security threats. Quantum computing can break traditional encryption methods, but it can also enhance security by generating unbreakable cryptographic keys. AI can improve threat detection and response through advanced analytics and automation [73]. AI is exclusively appropriate for the following tasks:

- Perceiving actual cyberattacks more accurately than humans.
- Creating fewer false-positive results and arranging responses based on their real-world risks.
- Recognizing and flagging the type of distrustful emails and messages often employed in phishing campaigns.
- Simulating social engineering attacks, which helps security teams spot potential vulnerabilities before cybercriminals exploit them.
- Analyzing vast amounts of incident-related to data rapidly, so that security teams can swiftly take action to contain the threat.

5.5.7 Abusing language detection in online user content

The words "abusing language" or abusive behavior represent a broad category of content that includes hate speech, profanity, threats, cyberbullying, and various ethnic and racial insults. Each category can be considered abusive, and these categories are not mutually exclusive. There is no single definition of abuse, and often terms like "harassment," "abusive language," and "harmful speech" have been used interchangeably. Online abusive behavior received much research consideration in recent years because of its prevalence and severe consequences reported in the media. Online abuse affects youths' self-esteem and academic performances, and leads to anxiety, depression, and suicidal ideation [74, 75].

5.5.8 Transportation and logistics

Quantum computing can improve routing and planning for vehicles and transportation networks. AI can enhance predictive maintenance and real-time traffic management. The growth of AI has transformed the field of logistics. Significant developments from this technology have been carried out in various ways, including autonomous vehicles, warehouse automation, and smart roads. This powerful technology will automate and streamline many processes, consequently helping companies save both money and time [76].

5.5.9 Climate modeling

It has been found that the computational complexity of the modern climate models leftover a central bottleneck in understanding and calculating regional climate change under various manmade emission situation. Quantum computing can simulate complex climate models and optimize energy systems. AI can help analyze satellite and sensor data for weather forecasting and disaster response. A machine learning approach can be utilized for a unique dataset of prevailing climate model simulations to learn relationships between short-term and long-term temperature responses to different climate scenarios [77, 78].

The applications of AI and quantum computing are enormous and influential with the prospective to influence all fields and the industry.

5.5.10 Deep learning for the design of photonic structures

Nanophotonics has had a thoughtful influence in the field of optics by empowerment and manipulation of the light and matter interaction with subwavelength structures. However, despite the many advances in this field, its influence and dissemination in our daily lives has been delayed by a complicated and iterative process, cycling through modeling, nanofabrication, and nano-characterization. The fundamental reason is the fact that not only is the prediction of the optical response very time-consuming, requiring that Maxwell's equations be solved with dedicated numerical packages [79–81], but also the inverse problem, i.e., designing a nanostructure with an on-demand optical response is presently an excessive task, even with advanced numerical tools and techniques due to the high nonlinearity of the problem. However, deep learning illustrates its capability to envisage the geometry of nanostructures based merely on their far-field response. This method reports a direct way to the currently unreachable reverse problem, breaking ground for on-demand design of optical response with applications such as imaging, sensing, and plasmons mediated cancer thermotherapy [82].

5.5.11 AI for quantum communication and error correction

Quantum communication is based on the fundamental principles of quantum mechanics, including principles of uncertainty, noncloning, quantum entanglement, and measurement collapse. Quantum communication is composed to catalyzc a new age of the technological revolution and transform the communication industry. One of the noticeable methods of quantum communication is quantum key distribution (QKD), which snugly shares a quantum key between distant locations via the transmission of quantum states and encrypts information one bit at a time using the key. It is clear that the integration of AI with the quantum communication system can improve the security and efficiency of QKD [83–85]. Quantum experiments indicate a vital step toward creating the quantum computers. For example, 3-particle photonic quantum states represent a building block for a photonic quantum computing architecture. In the reinforcement learning algorithm, an agent puts the optical elements on the optical table. Each stroke complements an optical element to an existing setup. If the resulting setup achieves the goal, e.g., creates a desired multiphoton entangled state, then the agent receives a reward. This type of learning system is illustrated in Figure 5.5(a). As an example, an experiment on an optical table is provided. The learning scheme shown in the figure depicts how an agent, the reinforcement learning algorithm, learns to design the quantum experiments. The reinforcement learning algorithm that attains the experimental designs is depicted in Figure 5.5(b). It represents the reinforcement learning algorithm, projective simulation, as a two-layered network of the clips. The first layer resembles the current state of an optical system, while the second layer is the layer of the actions. The connectives between the layers express the memory of agent, which alters during the learning process. The connectivities resemble the probabilities of reaching a definite action in the given state of a quantum optical system. During such a learning process, an agent robotically adjusts the connectivity, and hence arranges various actions different than the other.

The reinforcement learning method to quantum protocol design has been found to be applicable for a diversity of quantum communication tasks, including entanglement purification, quantum teleportation, and quantum repeater. The arrangement of the learning setting is revealed in Figure 5.6. The agent notices the quantum environment state and selects an act based on the projective simulation deliberation process. In other words, the algorithm chooses how to modify this state by acting on the environment.

5.5.12 AI for quantum control

The progression in the field of quantum control has the prospective to improve human understanding of the microcosm and normalize the behavior of the quantum systems. The quantum control system focuses on regulating

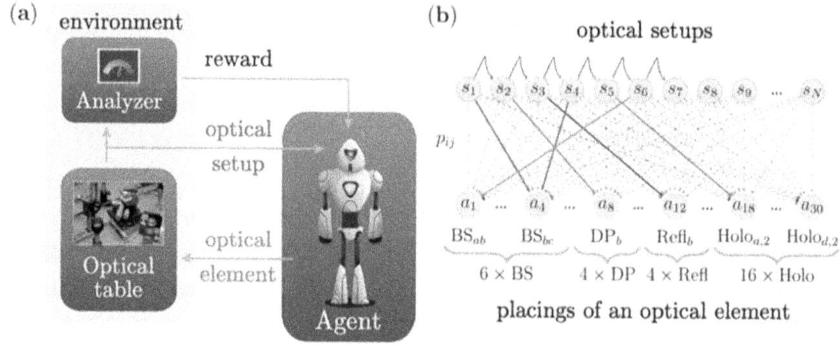

Figure 5.5 A reinforcement learning algorithm that designs a quantum experiment (adapted from [27]).

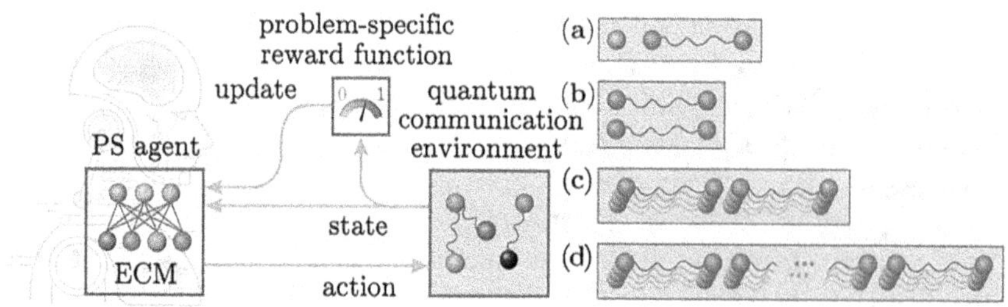

Figure 5.6 The schematic of a reinforcement learning algorithm that designs a long-distance quantum communication protocol. The algorithm is based on projective simulation with episodic and compositional memory (ECM). The goals of the planner/scheduler (PS) agent are (*a*) teleportation protocol, (*b*) entanglement purification protocol, (*c*) quantum repeater protocol, and (*d*) quantum repeater protocol for long-distance quantum communication (adapted from [27]).

the behavior of the micro quantum mechanical systems. The objective is to design and implement control through custom external fields so that the dynamics of quantum control systems can progress in a precise way. One notable section of a traditional quantum control device comprises measurement feedback methods. The schematic diagram of a measurement feedback method is presented in Figure 5.7. The quantum control system needs measurement, and state estimation can be carried out by employing the completeness of the observed quantities. This denotes the calculation of the complete state of the system by combining the probability of the quantum projection of the state on various mechanical quantities. Henceforth, the original matrix is recreated using the density matrix's projection in several directions. The implementation approach is then fed back to the controlled system based on the assessed system status.

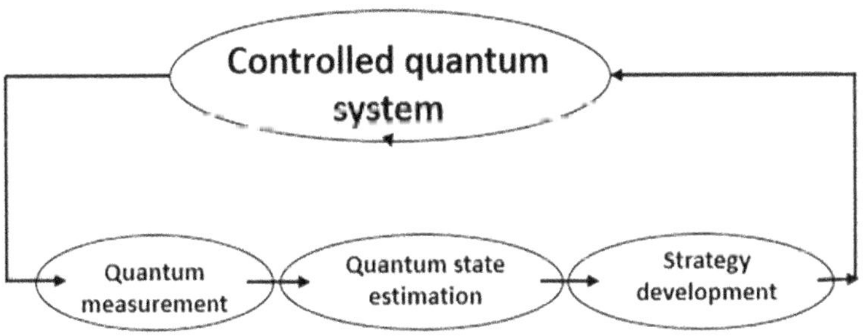

Figure 5.7 An architecture of the quantum control system based on the measurement feedback.

5.5.13 AI for quantum phases of matter

The classification capabilities of AI are widely known in the macroscopic world and are known to surpass human performance in certain contexts. Such capabilities are also being functionalized to quantum many body systems, predominantly in the identification of the quantum phases. The probability distribution of the quantum states follows a wave-like design, in which probability is estimated by knowing the amplitude of the wave. Furthermore, the phase of the wave carries the information. The role of the phase in quantum mechanics is very significant characteristic. In the reference of the time-dependent Hamiltonian, the potential field cultivates according to a periodic function of the time, which generates a wave function that is unable to return to its initial state after a complete period of evolution. In fact, the significance of the phase extends beyond the quantum mechanics, as it also plays a critical role in other areas of physics. At present, supervised learning is the best commonly employed classification algorithm for identification of the phase of the matter [86]. This type of algorithm enables the clear detection of the new phases, the localization of the phase bound arise and the accurate classification. Nonetheless, for the complex multibody systems with inadequate theoretical knowledge of simulation models, recognizing a new physics from a large amount of measured data can be grueling [87]. A hybrid method that combines unsupervised and supervised learning practices has also been developed as a capable strategy [88].

REFERENCES

[1] Sriarunothai T., et. al., Speeding-up the decision making of a learning agent using an ion trap quantum processor'. *Quantum Sci. Technol.* 2019;4:015014. https://doi.org/10.1088/2058-9565/aaef5e

[2] Saggio V., et. al., Experimental quantum speed-up in reinforcement learning agents'. *Nature* 2021;591:229.

[3] Liu Y.P., et. al. Quantum reinforcement learning method and application based on value function. *IFAC PapersOnLine* 2022;55–11:132–137.

[4] Dalla Pozza, N., Quantum reinforcement learning: the maze problem. *Quantum Machine Intelligence* 2022;4:11. https://doi.org/10.1007/s42484-022-00068-y
[5] Einstein A., Podolsky B., Rosen N. Can quantum-mechanical description of physical reality be considered complete. *Phys Rev.* 1935; 47:777–780.
[6] Aimeur E., Brassard G., Gambs S., Quantum clustering algorithms. In *ICML '07: Proceedings of the 24th international conference on Machine learning, Corvallis*, Oregon, USA; 2007. p. 1–8.
[7] Harrow A.W., Hassidim A., Lloyd S. Quantum algorithm for linear systems of equations. *Phys Rev Lett.* 2009; 103:150502.
[8] Lloyd S., Mohseni M., Rebentrost P. Quantum principal component analysis. *Nat Phys.* 2014; 10:631–633.
[9] Rebentrost P., Mohseni M., Lloyd S. Quantum support vector machine for big data classification. *Phys Rev Lett.* 2014;113: 130503.
[10] Farhi E., Goldstone J., Gutmann S. A. quantum approximate optimization algorithm. arXiv:1411.4028, 2014.
[11] Peruzzo A., McClean J., Shadbolt P., et al. A variational eigenvalue solver on a photonic quantum processor. *Nat Commun.* 2014;07:5:7.
[12] Krenn M., Malik M., Fickler R., et al. Automated search for new quantum experiments. *Phys Rev Lett.* 2016; 116:090405.
[13] Melnikov A.A., Poulsen Nautrup H., Krenn M., et al. Active learning machine learns to create new quantum experiments. *Proc Natl Acad Sci USA.* 2018; 115:1221–1226.
[14] Farhi E., Neven H. Classification with quantum neural networks on near term processors. arXiv:1802.06002, 2018.
[15] Cong I., Choi S., Lukin M.D. Quantum convolutional neural networks. *Nat Phys.* 2019; 15: 1273–1278.
[16] Fosel T., Tighineanu P., Weiss T., et al. Reinforcement learning with neural networks for quantum feedback. *Phys Rev X.* 2018; 8:031084.
[17] Poulsen Nautrup H., Delfosse N., Dunjko V., et al. Optimizing quantum error correction codes with reinforcement learning. *Quantum.* 2019; 3:215.
[18] Melnikov A.A., Fedichkin L.E., Alodjants A. Predicting quantum advantage by quantum walk with convolutional neural networks. *New J Phys.* 2019; 21:125002.
[19] Melnikov A.A., Fedichkin L.E., Lee R-K., et al. Machine learning transfer efficiencies for noisy quantum walks. *Adv Quantum Technol.* 2020; 3: 1900115.
[20] Jerbi S., Gyurik C., Marshall S., et al. Parametrized quantum policies for reinforcement learning. *Adv Neural Inf Process Syst.* 2021; 34:28362–28375.
[21] Saggio V., Asenbeck B.E., Hamann A., et al. Experimental quantum speed-up in reinforcement learning agents. *Nature.* 2021; 591:229–233, 03.
[22] Perelshtein M., Sagingalieva A., Pinto K., et al. Practical application-specific advantage through hybrid quantum computing. arXiv:2205.04858, 2022.
[23] Sagingalieva A., Kurkin A., Melnikov A., et al. Hyperparameter optimization of hybrid quantum neural networks for car classification. arXiv:2205.04878, 2022.
[24] Sagingalieva A., Kordzanganeh M., Kenbayev N., et al. Hybrid quantum neural network for drug response prediction. arXiv:2211.05777, 2022.
[25] You Y., Lai X., Pan Y. et al. Artificial intelligence in cancer target identification and drug discovery. *Sig Transduct Target Ther.* 2022;7:156. https://doi.org/10.1038/s41392-022-00994-0

[26] Shih, M, Quillin H. D. N., Garrett K. E., Coyle E. J., Secondo R., Keyser C. K., Mills M. S., Harper E. S., Maximizing supercontinuum bandwidths in gas-filled hollowcore fibers using artificial neural networks. *J. Appl. Phy*. 2023; 133:233101.

[27] Melnikov A, Quantum machine learning: from physics to software engineering. *Advances in Physics*. 2023;X 8"2165452.

[28] Sekatski P., Bancal J.-D., Valcarce X., et al. Device-independent quantum key distribution from generalized CHSH inequalities. *Quantum*. 2021; 5:444.

[29] Canabarro A., Brito S., Chaves R. Machine learning nonlocal correlations. *Phys Rev Lett*. 2019; 122: 200401.

[30] Melnikov A.A., Sekatski P., Sangouard N. Setting up experimental Bell tests with reinforcement learning. *Phys Rev Lett*. 2020; 125:160401.

[31] Zhu Y., Yu K. Artifcial intelligence (AI) for quantum and quantum for AI. *Optical and Quantum Electronics*. 2023;55:697.

[32] Chen M.K., Liu X., Sun Y., Tsai D. P. Artificial intelligence in meta-optics. *Chem. Rev*. 2022;122:15356–15413.

[33] Park J.-S., Zhang S., She A., Chen W. T., Lin P., Yousef K. M., Cheng J.-X., Capasso F. All-glass, large metalens at visible wavelength using deep-ultraviolet projection lithography. *Nano Lett*. 2019;19:8673–8682.

[34] Tseng M. L., Lin Z. H., Kuo H. Y., Huang T. T., Huang Y. T., Chung T. L., Chu C. H., Huang J. S., Tsai D. P. Stress-induced 3D chiral fractal metasurface for enhanced and stabilized broadband near-field optical chirality. *Adv. Opt. Mater*. 2019;7:1900617.

[35] Hadibrata W., Wei H., Krishnaswamy S., Aydin K. Inverse design and 3D printing of a metalens on an optical fiber tip for direct laser lithography. *Nano Lett*. 2021;21:2422–2428.

[36] Li Y., Li X., Chen L., Pu M., Jin J., Hong M., Luo X. Orbital angular momentum multiplexing and demultiplexing by a single metasurface. *Adv. Opt. Mater*. 2017;5:1600502.

[37] Zhang L., Chen X. Q., Liu S., Zhang Q., Zhao J., Dai J. Y., Bai G. D., Wan X., Cheng Q., Castaldi G., et al. Space-time-coding digital metasurfaces. *Nat. Commun*. 2018;9:4334.

[38] Yan L., Zhu W., Wu P. C., Cai H., Gu Y., Chin L. K., Shen Z., Chong P., Yang Z., Ser W., et al. Adaptable metasurface for dynamic anomalous reflection. *Appl. Phys. Lett*. 2017;110:201904.

[39] Wu P. C., Tsai W. Y., Chen W. T., Huang Y. W., Chen T. Y., Chen J. W., Liao C. Y., Chu C. H., Sun G., Tsai D. P. Versatile polarization generation with an aluminum plasmonic metasurface. *Nano Lett*. 2017;17:445–452.

[40] Wu P. C., Chen J. W., Yin C. W., Lai Y. C., Chung T. L., Liao C. Y., Chen B. H., Lee K. W., Chuang C. J., Wang C. M., et al. Visible metasurfaces for on-chip polarimetry. *ACS Photonics*, 2018;5:2568–2573.

[41] Yan L., Zhu W., Karim M. F., Cai H., Gu A. Y., Shen Z., Chong P. H. J., Tsai D. P., Kwong D. L., Qiu C. W., et al. Arbitrary and independent polarization control in situ via a single metasurface. *Adv. Opt. Mater*. 2018;6:1800728.

[42] Frese D., Wei Q., Wang Y., Cinchetti M., Huang L., Zentgraf T. Nonlinear bicolor holography using plasmonic metasurfaces. *ACS Photonics*, 2021;8:1013–1019.

[43] Jin L., Huang Y.-W., Jin Z., Devlin R. C., Dong Z., Mei S., Jiang M., Chen W. T., Wei Z., Liu H., et al. Dielectric multi-momentum meta-transformer in the visible. *Nat. Commun*., 2019;10:4789.

[44] Saini T.S., Tuan T.H., Matsumoto M., Sakai G., Suzuki T., Ohishi Y., Mid-infrared wavelength conversion using dispersion engineered As2S5 micro-structured optical fiber pumped with ultrafast laser at 2 µm. *Opt. Letters* 2020;45(10):2736–2739.

[45] Saini T.S., Sinha R.K., Mid-infrared supercontinuum generation in soft-glass specialty optical fibers: a review. *Progress in Quantum Electronics*, 2021;78:100342.

[46] Saini T.S., Arora S., Supradeepa V.R., Far-detuned mid-IR wavelength conversion at 4.05 µm in a tellurium oxide rib waveguide pumped at 1550 nm: design and analysis. *AIP Advances* 2021;11:055110.

[47] Saini T.S., Supradeepa V.R., Tellurium-oxide coated silicon-nitride hybrid waveguide for near-to-mid-IR supercontinuum generation: design and analysis. *J. Mod. Opt.* 2021;68(1):29–36.

[48] Zhang C., Divitt S., Fan Q., Zhu W., Agrawal A., Lu Y., Xu T., Lezec H. J. Low-loss metasurface optics down to the deep ultraviolet region. *Light Sci. Appl.* 2020;9:55.

[49] Shen K.-C., Huang Y.-T., Chung T. L., Tseng M. L., Tsai W. Y., Sun G., Tsai D. P. Giant efficiency of visible second-harmonic light by an all-dielectric multiple-quantum-well metasurface. *Phys. Rev. Appl.* 2019;12:064056.

[50] Shen K. C., Ku C. T., Hsieh C., Kuo H. C., Cheng Y. J., Tsai D. P. Deep-ultraviolet hyperbolic metacavity laser. *Adv. Mater.* 2018;30:1706918.

[51] Azzam S. I., Chaudhuri K., Lagutchev A., Jacob Z., Kim Y. L., Shalaev V. M., Boltasseva A., Kildishev A. V. Single and multi-mode directional lasing from arrays of dielectric nanoresonators. *Laser Photonics Rev.* 2021;15:2000411.

[52] Qu G., Yang W., Song Q., Liu Y., Qiu C.-W., Han J., Tsai D. P., Xiao S. Reprogrammable meta-hologram for optical encryption. *Nat. Commun.* 2020; 11:5484.

[53] Nemati A., Wang Q., Ang N. S. S., Wang W., Hong M., Teng J. Ultra-high extinction-ratio light modulation by electrically tunable metasurface using dual epsilon-near-zero resonances. *Opto-Electron. Adv.* 2021;4:200088.

[54] Zwolak J.P., et. al., Auto-tuning of double dot devices in situ with machine learning. *Physical Review Applied*, 2020;13:034075.

[55] Li J., Yu P., Zhang S., Liu N. Electrically-controlled digital metasurface device for light projection displays. *Nat. Commun.* 2020;11:3574.

[56] Yang J., Ghimire I., Wu P. C., Gurung S., Arndt C., Tsai D. P., Lee H. W. H. Photonic crystal fiber metalens. *Nanophotonics*, 2019;8:443–449.

[57] Shalaginov M. Y., An S., Yang F., Su P., Lyzwa D., Agarwal A. M., Zhang H., Hu J., Gu T. Single-element diffraction-limited fisheye metalens. *Nano Lett.* 2020;20:7429–7437.

[58] Wesemann L., Davis T. J., Roberts A. Meta-optical and thin film devices for all-optical information processing. *Appl. Phys. Rev.*, 2021;8:031309.

[59] Huang K., Liu H., Restuccia S., Mehmood M. Q., Mei S.-T., Giovannini D., Danner A., Padgett M. J., Teng J.-H., Qiu C.-W. Spiniform phase-encoded metagratings entangling arbitrary rationalorder orbital angular momentum. *Light Sci. Appl.*, 2018;7:17156–17156.

[60] Chen B. H., Wu P. C., Su V.-C., Lai Y.-C., Chu C. H., Lee I. C., Chen J.-W., Chen Y. H., Lan Y.-C., Kuan C.-H., et al. GaN metalens for pixel-level full-color routing at visible light. *Nano Lett.*, 2017;17:6345–6352.

[61] Chen M. K., Yan Y., Liu X., Wu Y., Zhang J., Yuan J., Zhang Z., Tsai D. P. Edge detection with meta-lens: from dimension to three dimensions. *Nanophotonics*, 2021;10:3709 3715.
[62] Zheng P., Dai Q., Li Z., Ye Z., Xiong J., Liu H.C., Zheng G., Zhang S. Metasurface-based key for computational imaging encryption. *Sci. Adv.* 2021;7:No. eabg0363.
[63] Faraji-Dana M., Arbabi E., Arbabi A., Kamali S. M., Kwon H., Faraon A. Compact folded metasurface spectrometer. *Nat. Commun.*, 2018;9:4196.
[64] Zhao M., Chen M. K., Zhuang Z.-P., Zhang Y., Chen A., Chen Q., Liu W., Wang J., Chen Z.-M., Wang B., et al. Phase characterisation of metalenses. *Light Sci. Appl.* 2021;10:52.
[65] Tsai D.-P. Exploring the electromagnetic information of metasurfaces. *Natl. Sci. Rev.* 2020;7:1845–1846.
[66] Li Z., Lin P., Huang Y.-W., Park J.-S., Chen W. T., Shi Z., Qiu C.-W., Cheng J.-X., Capasso F. Meta-optics achieves RGB achromatic focusing for virtual reality. *Sci. Adv.* 2021;7:No. eabe4458.
[67] Zhou J., Liu S., Qian H., Li Y., Luo H., Wen S., Zhou Z., Guo G., Shi B., Liu Z. Metasurface enabled quantum edge detection. *Sci. Adv.* 2020;6:No. eabc4385.
[68] Davenport T., Kalakota R., The potential for artificial intelligence in healthcare. *Future Healthcare Journal* 2019;6(2):94–98.
[69] Sajjan M., Junxu L, Selvarajan R., et al. Quantum machine learning for chemistry and physics. *Chem Soc Rev.* 2022;51:6475.
[70] Hosny A., Parmar C., Quackenbush J., Schwartz L.H., Aerts H.J.W.L., Artificial intelligence in radiology. *Nature Reviews Cancer* 2018;18:500–510.
[71] Hasselgren C., Operea T. I., Artificial intelligence for drug discovery: Are we there yet? *Annu. Rev. Pharmacol. Toxicol.* 2024;64:527–550.
[72] Goodell J. W., Kumar S., Lim W. M., Pattnaik D., Artificial intelligence and machine learning in finance: Identifying foundations, themes, and research clusters from bibliometric analysis. *J. Behav. Exp. Finance* 2021;32:100577.
[73] Awad W.A., Elseuofi S.M., Machine learning methods for spam email classification. *Int. J. Comp. Sci. Info. Technol.* 2011;3(1):173.
[74] Kaur S., Singh S., Kaushal S., Abusive content detection in online user-generated data: a survey. *Procedia Computer Science* 2021;189:274–281.
[75] Wang K., Lu D., Han S.C., Long S., Poon J., Detect all abuse! toward universal abusive language detection models. *Proc. 28th Intern. Conf. Computational Linguistics*, 6366–6376, Barcelona, Spain (Online), December 8–13, 2020.
[76] https://www.ilscompany.com/the-role-of-artificial-intelligence-in-logistics/
[77] Mansfield et al. Predicting global patterns of long-term climate change from short-term simulations using machine learning. *NPJ Climate and Atmospheric Science* 2020;3:44.
[78] Schneider T., Behera S., Boccaletti G. et al. Harnessing AI and computing to advance climate modelling and prediction. *Nat. Clim. Chang.* 2023;13:887–889.
[79] Multiphysics C., "v. 5.2," COMSOL AB, Stockholm, Sweden, 2015, www.comsol.com
[80] Oskooi A.F., Roundy D., Ibanescu M., Bermel P., Joannopoulos J.D., Johnson S.G., Meep: A flexible free software package for electromagnetic simulations by the FDTD method. *Computer Physics Communications*, 2010;181(3):687–702.
[81] Solutions F., Lumerical solutions. Inc., https://www.lumerical.com, 2003.

[82] Wei M, Liu Z, Kudyshev ZA, et al. Deep learning for the design of photonic structures. *Nat Photonics* 2021;15:77.
[83] Wang W., Lo H.-K. Machine learning for optimal parameter prediction in quantum key distribution. *Phys. Rev. A* 2019;100(6), 62334.
[84] Liu J.-Y., et al. Practical phase-modulation stabilization in quantum key distribution via machine learning. *Phys. Rev. Appl.* 2019;12(1):14059.
[85] Liu Z.-P., et al. Automated ML for secure key rate in discrete-modulated continuous-variable quantum key distribution. *Opt. Exp.* 2022;30(9): 15024–15036.
[86] Zhao E., et al. Observing a topological phase transition with deep neural networks from experimental images of ultracold atoms. *Opt. Exp.* 2022;30(21):37786–37794.
[87] Lidiak A., Gong Z., Unsupervised machine learning of quantum phase transitions using diffusion maps. *Phys. Rev. Lett.* 2020;125(22):225701.
[88] Miles C., et al. Machine learning discovery of new phases in programmable quantum simulator snapshots. *Phys. Rev. Res.* 2023;5(1):13026.

Chapter 6

Deep learning–enabled optical sensors

Gargi Srivastava and Sajal Agarwal

6.1 INTRODUCTION TO DEEP LEARNING AND OPTICAL SENSORS

Optical sensing is a powerful technology that uses light to gather information about the properties of objects, materials, or environments. It relies on the interaction of light with the target, capturing and analyzing the light signals that are reflected, transmitted, or scattered [1]. The basic components of an optical sensing system include a light source, a sensing element (e.g., photodetector), and signal processing mechanisms. Optical sensors can be of many types, and the analysis parameter will be different based on the type. Optical sensors can be realized using fiber-based structure and/or using chip-based structure [2]. There are three inherent parts of any optical sensing system, namely, transmitter, medium, and receiver. Optical sensing requires a light source that emits light of specific wavelengths. Light sources can be LEDs (light-emitting diodes), lasers, or broadband sources like incandescent bulbs or supercontinuum sources [3]. Photodetectors are essential components of optical sensing systems. They convert light signals into electrical signals. Common types of photodetectors include photodiodes, phototransistors, and photomultiplier tubes (PMTs). Various sensing mechanisms are employed in optical sensors, depending on the application. Examples include absorption spectroscopy, fluorescence, interferometry, and Raman scattering [4–6]. Optical fibers play a crucial role in guiding and transmitting light signals between the light source, target, and photodetector. They offer flexibility and allow remote sensing. Signal processing techniques are used to extract meaningful information from the collected optical data. This can involve filtering, Fourier analysis, or other advanced algorithms [7].

The significance of optical sensors for sensing lies in their ability to harness the unique properties of light to noninvasively and accurately measure and analyze a wide range of parameters in diverse applications. Optical sensors offer high sensitivity, enabling detection of subtle changes in target properties, and their nondestructive nature makes them ideal for sensitive environments, such as medical diagnostics and biological research [8]. With

DOI: 10.1201/9781003591030-6

fast response times, optical sensors provide real-time data for rapid decision-making in critical processes, like industrial monitoring and quality control [9]. Additionally, optical sensors can cover long detection ranges, making them indispensable for remote sensing and environmental monitoring. Their multiplexing capabilities further enhance their versatility, enabling simultaneous measurement of multiple parameters with cost-effective solutions [10]. The significance of optical sensors lies in their broad applications, precision, and adaptability, which drive advancements in various fields and contribute to innovation and progress.

Optical sensing offers a multitude of advantages across various domains; it can provide even more powerful and sophisticated solutions for data analysis and decision-making in complex environments [8–10]. The future of optical sensing holds tremendous potential for advancements in technology and innovation due to its easy integrability with deep learning. However, it is important to note that integrating deep learning with optical sensors requires large and diverse datasets for training the models effectively [11]. Additionally, considerations for power consumption and computational requirements should be taken into account, especially for portable or low-power sensor applications.

Machine learning and deep learning have revolutionized the field of optical sensing, offering unprecedented opportunities and advancements in this domain. The significance of machine learning and deep learning for optical sensing lies in their capacity to extract meaningful patterns, features, and relationships from vast and complex optical data [11]. By leveraging these advanced algorithms, optical sensors can achieve enhanced accuracy, sensitivity, and robustness in detecting and interpreting signals, leading to improved measurements and analysis. Machine learning models can be trained to identify specific targets or anomalies in real time, enabling rapid and automated decision-making in critical applications like environmental monitoring or health-care diagnostics [8]. Deep learning techniques, with their ability to handle large datasets and learn hierarchical representations, further augment optical sensing capabilities by reducing noise and improving classification accuracy, which is vital in fields such as material identification and chemical analysis [12]. Moreover, machine learning and deep learning allow for self-calibration and adaptability, ensuring continuous and precise operation of optical sensing systems in dynamic environments. With the integration of these technologies, optical sensors become more versatile, and cost-effective, and offer promising solutions to a wide range of challenges, making them indispensable tools for scientific research, industry, and various societal applications [13].

6.2 FUNDAMENTALS OF OPTICAL SENSOR TECHNOLOGY

Optical sensors are broadly divided into three categories.

6.2.1 Through-beam sensors

Through-beam optical sensors, also known as "photoelectric sensors," operate on the principle of interrupting a continuous beam of light between a transmitter and a receiver to detect the presence or absence of an object. The sensor consists of two main components: a light transmitter and a light receiver placed opposite each other [14]. The transmitter emits a constant beam of light, typically infrared, toward the receiver. When an object enters the sensor's detection zone and interrupts the beam, the receiver detects the absence of light and triggers an output signal, indicating the presence of the object. This technology is widely used in various industrial applications such as packaging, conveyor systems, and automation due to its high accuracy, fast response time, and ability to detect even small objects [15].

6.2.2 Retro-reflective sensors

Retro-reflective optical sensors operate on the principle of utilizing a reflective target to bounce light back to the sensor for object detection. The sensor system comprises a light emitter and a combined receiver and detector. The emitter emits a continuous beam of light, often in the infrared spectrum, toward a reflector positioned opposite the sensor. The reflector is designed to redirect the emitted light back to the sensor's combined receiver and detector. In the absence of an object within the sensor's field of view, the reflected light precisely aligns with the receiver, resulting in a baseline signal level [16]. When an object enters the detection zone and disrupts the reflected light path, the receiver registers a decrease in the received light intensity, prompting the detector to trigger an output signal indicating the presence of the object. Retro-reflective sensors offer advantages in applications where detecting the presence or absence of objects is critical, such as in automatic door systems, conveyor belt control, and object counting. Their ability to reliably sense objects regardless of color or surface characteristics, coupled with the convenience of using a single unit for both transmission and reception, makes them an efficient choice for industrial automation. This sensor type can be equipped with adjustable sensitivity settings and precise alignment features to optimize performance based on specific application requirements [17].

6.2.3 Diffuse reflection sensors

Diffuse reflection optical sensors, also known as proximity sensors, operate based on the principle of detecting the presence of an object by measuring the amount of light reflected off the object's surface. These sensors consist of a light emitter and a separate light receiver placed side by side. The emitter emits a beam of light, typically in the form of infrared radiation, that illuminates the nearby area [18]. When an object enters the detection zone of the sensor, it reflects a portion of the emitted light back toward the receiver.

The receiver detects the intensity of the reflected light and analyzes it. The key factor that determines object detection is the intensity of the reflected light. If the object is close to the sensor, a higher intensity of light will be reflected back, indicating the presence of an object. Conversely, if the object is distant or not present, the intensity of the reflected light will be lower. To ensure accurate detection and reliable operation, diffuse reflection sensors often incorporate various features such as adjustable sensitivity settings and threshold adjustments. These allow the sensor's response to be fine-tuned for different objects or environmental conditions [19]. Additionally, some sensors utilize background suppression techniques to mitigate interference from reflective surfaces other than the target object. Diffuse reflection sensors find applications in a wide range of industries, including manufacturing, packaging, robotics, and automotive. They are particularly useful for tasks such as object counting, presence detection, and part positioning. Figure 6.1 has the source and receiver arrangement for various optical sensors. The comparative analysis of various sensor is shown in Table 6.1.

From the discussion, it is believed that integration of machine learning with optical sensors is doable and can lead to enhanced capabilities and improved performance in various applications. The combination of machine learning algorithms and optical sensor data can provide valuable insights, automation, and adaptive functionalities that traditional sensor systems alone might struggle to achieve.

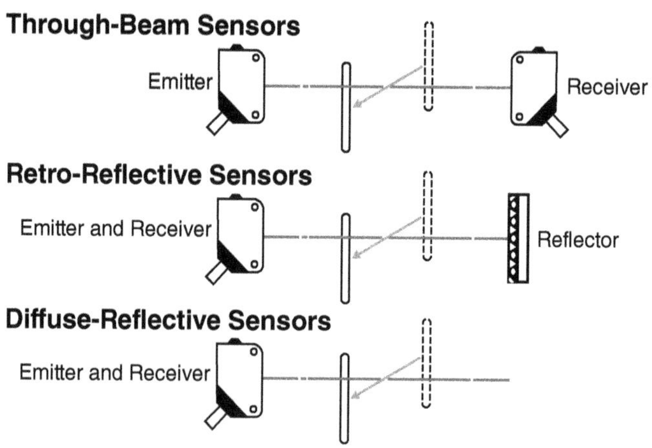

Figure 6.1 Arrangement of optical source and receiver for various optical sensors [20].

Table 6.1 Comparative analysis of various optical sensors

Aspect	*Through-beam optical sensor*	*Retro-reflective optical sensor*	*Diffuse reflection optical sensor*
Working Principle	Emits light from transmitter and detects interruption in received light at separate receiver	Emits light toward a reflector and detects reflection from a target	Emits light and detects intensity of light reflected from target surface
Setup Complexity	More complex setup due to separate emitter and receiver units	Moderately complex setup with emitter, reflector, and receiver	Simplified setup with integrated emitter and receiver
Detection Range	Longest detection range among the three types	Moderate detection range, shorter than through-beam	Limited detection range, typically shorter than other types
Interference	Less susceptible to background interference	Susceptible to interference from other reflective surfaces	Prone to interference from ambient light and reflective surfaces
Object Characteristics	Can detect objects of various sizes and materials	Suited for detecting objects with a reflective surface	Can detect objects based on reflectivity, suitable for nonreflective objects
Applications	Used in applications requiring long-distance detection and precise alignment	Suitable for applications where reflectors can be installed, such as door control	Commonly used for proximity sensing, object counting, and part presence detection
Ease of Installation	Requires careful alignment of separate emitter and receiver	Requires proper alignment of reflector and sensor	Relatively easy to install and align
Cost	Moderate cost due to separate units	Reflectors add to the cost, moderate pricing	Generally, more cost-effective
Examples	Automatic door systems, precise object detection	Conveyor belt control, presence detection	Object counting, packaging lines, robotics

6.3 DEEP LEARNING TECHNIQUES FOR OPTICAL SENSORS

Deep learning algorithms have gained remarkable attention in recent years due to their exceptional performance in various applications involving complex data. Optical sensor data, which encompasses information captured through cameras, LiDAR, and other optical sensors, presents a unique set of challenges and opportunities for deep learning techniques. This section provides an overview of key deep learning algorithms that are relevant to processing and extracting insights from optical sensor data.

Convolutional neural networks (CNNs) have revolutionized image analysis tasks by automatically learning hierarchical features from raw pixel data. CNNs employ convolutional layers to detect local patterns, followed by pooling layers for downsampling and reducing spatial dimensions [21]. These architectures are particularly well-suited for optical sensor data like images and videos, enabling tasks such as object detection, classification, and semantic segmentation.

Recurrent neural networks (RNNs), with their ability to handle sequential data, find relevance in optical sensor data captured over time. Long short-term memory (LSTM) networks, a type of RNN, can capture temporal dependencies and patterns in time-series optical sensor data [22]. This makes them valuable for tasks like gesture recognition, action prediction, and video analysis.

Generative adversarial networks (GANs) introduce a novel approach to data generation [23]. They consist of a generator network that generates data samples and a discriminator network that distinguishes between real and generated samples. GANs have been used for tasks like image-to-image translation, super-resolution, and data augmentation for optical sensor data.

Attention mechanisms have proven highly effective in focusing on relevant parts of input data, which is crucial for processing large and complex optical sensor data [24]. Transformer-based architectures, known for their attention mechanisms, have led to breakthroughs in tasks like image captioning, optical character recognition, and image synthesis.

Transfer learning involves utilizing pretrained deep learning models on large datasets and adapting them to specific tasks with limited data [25]. This is particularly valuable for optical sensor data, where labeled data might be scarce. Transfer learning enables faster model convergence and improved generalization.

Deep learning algorithms have ushered in a new era of possibilities for processing and analyzing optical sensor data. CNNs, RNNs, GANs, attention mechanisms, and transfer learning techniques have all contributed to significant advancements in various optical sensor-related applications. As research in deep learning continues to evolve, we can anticipate even more sophisticated algorithms tailored to the intricacies of optical sensor data, enabling us to unlock valuable insights from the visual world.

6.3.1 Deep neural networks for signal processing and feature extraction

Signal processing and feature extraction are fundamental tasks in various fields, ranging from telecommunications to image analysis and beyond. Deep neural networks (DNNs) have emerged as a powerful paradigm for addressing these tasks. With their ability to automatically learn intricate patterns and representations from raw data, DNNs have revolutionized signal processing and feature extraction, leading to improved performance and expanded capabilities across multiple domains. This section explores the significance of DNNs in signal processing and feature extraction and discusses key reference works in this area.

Traditional signal processing methods often require manual engineering of features, a process that can be time-consuming and reliant on domain expertise. DNNs, on the other hand, can autonomously learn relevant features and representations from raw data, bypassing the need for explicit feature engineering. This capability has proven particularly valuable in fields like audio analysis, where DNNs can learn to differentiate between different speech phonemes or detect anomalies in acoustic signals.

CNNs have garnered substantial attention for their excellence in image analysis tasks [21]. CNNs utilize convolutional layers to extract hierarchical features, capturing both low-level patterns like edges and high-level structures such as object shapes. This architecture has revolutionized image classification, object detection, and semantic segmentation, finding applications in diverse fields such as medical imaging and autonomous driving.

RNNs are tailored for sequential data, making them apt for time-series analysis, natural language processing, and more [22]. RNNs can capture temporal dependencies, crucial for tasks like speech recognition or financial time-series prediction. Additionally, LSTM networks, a variant of RNNs, excel in modeling long-range dependencies, making them suitable for applications like language translation and music generation.

Autoencoders are unsupervised DNNs used for feature learning and data compression [26]. They consist of an encoder that maps input data to a latent space representation, and a decoder that reconstructs the input from this representation. Autoencoders have applications in anomaly detection, denoising, and dimensionality reduction, aiding in preprocessing data for downstream tasks.

Attention mechanisms have elevated feature extraction capabilities by enabling models to focus on relevant parts of the input [24]. These mechanisms have been successfully employed in tasks such as image captioning, machine translation, and sentiment analysis. Self-attention mechanisms, notably introduced by the transformer model, have redefined the state of the art in natural language processing and image analysis.

DNNs have reshaped the landscape of signal processing and feature extraction. Their capability to learn intricate patterns from raw data, coupled with architectures like CNNs for image analysis, RNNs for sequential data, autoencoders for unsupervised learning, and attention mechanisms for enhanced focus, has led to substantial advancements in multiple domains. These advancements not only improve accuracy and efficiency but also open doors to new applications and insights in fields as diverse as health care, finance, and communication.

6.4 DEEP LEARNING MODELS FOR OPTICAL SENSOR DATA ANALYSIS

6.4.1 CNNs for image-based optical sensors

In the age of information, the exponential growth of image data has led to an increasing demand for efficient and accurate ways to process and extract meaningful information from visual content. CNNs have emerged as a game-changer in the realm of image analysis, particularly for applications involving optical sensors. This section delves into the significance of CNNs in image-based optical sensor data analysis, highlighting their architecture, applications, and providing references for further exploration.

CNNs, inspired by the visual processing mechanisms of the human brain, have revolutionized the field of image analysis. Unlike traditional image processing techniques that rely on handcrafted features, CNNs automatically learn hierarchical features from raw pixel data. This ability to capture both low-level details and high-level structures has propelled CNNs into the spotlight for analyzing data captured by optical sensors.

CNNs consist of multiple layers designed to progressively extract and transform features. Convolutional layers apply filters to input images, detecting local patterns such as edges and textures. Pooling layers reduce spatial dimensions while retaining essential information. Fully connected layers learn complex relationships in the feature space. CNN architectures vary, with popular ones including LeNet, AlexNet, VGG, GoogLeNet, and ResNet, each designed to tackle specific image analysis tasks.

CNNs find extensive application in analyzing data from optical sensors, which encompass a range of devices like cameras, LiDAR, and hyperspectral imagery [27]. Some notable applications include the following:

1. **Object detection and recognition:** CNNs excel in detecting and recognizing objects within images captured by optical sensors [28]. This capability is invaluable in fields like autonomous driving, where CNN-based models identify pedestrians, vehicles, and road signs.

2. **Image classification:** CNNs are employed to classify images into predefined categories [29]. In the context of optical sensors, this has applications in quality control, sorting, and identification tasks across industries.
3. **Semantic segmentation:** Semantic segmentation involves classifying each pixel of an image into specific categories. CNNs enable accurate and real-time semantic segmentation of images from optical sensors, benefiting fields like medical imaging and agriculture.
4. **Anomaly detection:** CNNs learn the normal patterns present in optical sensor data, making them effective tools for detecting anomalies or defects in images. This is crucial in industrial inspection and surveillance applications.

CNNs have emerged as indispensable tools for processing and extracting insights from image-based optical sensor data. Their ability to automatically learn relevant features and hierarchies from raw pixel data has transformed image analysis tasks, enabling accurate object detection, image classification, semantic segmentation, and anomaly detection. As CNN architectures evolve and continue to demonstrate remarkable performance, they are poised to shape the future of image analysis across industries and applications.

6.4.2 RNNs for time-series optical sensor data

The rapid evolution of sensor technologies has led to an exponential growth in time-series data generated by various sensors, including optical sensors [30]. Extracting valuable insights from these temporal data streams is crucial across domains such as environmental monitoring, industrial automation, and health care. RNNs have emerged as a powerful tool for analyzing time-series data, allowing for the capture of temporal dependencies and patterns. This section delves into the significance of RNNs in processing time-series optical sensor data, discussing their architecture, applications, and providing references for further exploration.

Unlike traditional feedforward neural networks, RNNs are designed to handle sequential data by incorporating loops within their architecture, allowing information to persist across different time steps [31]. This makes RNNs ideal for modeling and predicting time-dependent patterns found in time-series data. Time-series optical sensor data typically involves measurements taken at consecutive time intervals, such as changes in light intensity, temperature variations, or motion patterns.

The core component of an RNN is the hidden state, which captures the network's memory of past information [32]. This memory is updated and influenced by the current input and the previous hidden state. The architecture of RNNs allows them to effectively process sequences of varying lengths. However, traditional RNNs suffer from the "vanishing gradient" problem, where gradients diminish over time, leading to difficulties in learning long-range dependencies.

RNNs offer a wide range of applications for processing time-series optical sensor data:

1. **Predictive maintenance:** RNNs can predict equipment failures by analyzing patterns in time-series data from sensors. This is crucial for avoiding costly downtime and optimizing maintenance schedules in industrial settings.
2. **Gesture recognition:** In the field of human-computer interaction, RNNs can recognize gestures based on time-series optical sensor data. This has applications in gaming, virtual reality, and health-care devices.
3. **Environmental monitoring:** RNNs can analyze time-series optical sensor data to predict pollution levels, weather changes, and other environmental variables, aiding in resource management and disaster preparedness.
4. **Health-care monitoring:** RNNs can process time-series data from wearable optical sensors to monitor vital signs, detect anomalies, and provide early warnings in medical emergencies.

Time-series optical sensor data is rich in temporal information, presenting both challenges and opportunities for analysis. RNNs have emerged as a potent solution for extracting patterns and dependencies from these dynamic data streams. By effectively capturing temporal relationships, RNNs empower applications such as predictive maintenance, gesture recognition, environmental monitoring, and health-care management. As RNN architectures continue to evolve and address challenges like vanishing gradients, their role in processing time-series optical sensor data is set to expand, contributing to advancements across numerous industries.

6.4.3 Other deep learning architectures for optical sensor analysis

The convergence of deep learning and optical sensor technology has given rise to transformative solutions across various domains. While CNNs and RNNs have been foundational in optical sensor analysis, other advanced deep learning architectures have emerged to tackle more complex challenges. This section delves into some of these innovative architectures, discussing their applications in optical sensor analysis and providing references for further exploration.

Gated recurrent units (GRUs) are a variant of RNNs that address the vanishing gradient problem more effectively [33]. GRUs utilize gating mechanisms to control the flow of information through the network, allowing them to capture long-range dependencies in time-series data. These architectures are particularly suitable for tasks involving temporal data from optical sensors, such as gesture recognition and speech processing.

Transformer architectures, originally designed for natural language processing, have demonstrated exceptional capabilities in various domains, including optical sensor analysis [24]. These architectures leverage self-attention mechanisms to capture relationships between different elements in a sequence, making them well-suited for tasks like sequence-to-sequence prediction, image captioning, and anomaly detection in optical sensor data.

Capsule networks (CapsNets) are a relatively recent architecture introduced as an alternative to CNNs for image analysis [34]. They aim to address some limitations of CNNs, such as their inability to efficiently handle spatial hierarchies and pose variations. CapsNets use "capsules" to represent specific features in an image, enabling better generalization and robustness in optical sensor data analysis.

Variational autoencoders (VAEs) are generative models that can learn meaningful latent representations of data [35]. They have applications in image generation, anomaly detection, and data compression. In optical sensor analysis, VAEs can capture complex patterns and variations, making them useful for tasks like anomaly detection in industrial environments or image synthesis.

As the complexity of optical sensor data analysis increases, advanced deep learning architectures have emerged as indispensable tools. GRUs, transformer architectures, CapsNets, and VAEs exemplify this evolution, addressing specific challenges in temporal data analysis, sequence prediction, feature representation, and generative modeling. These architectures expand the horizons of optical sensor analysis, enabling more accurate, robust, and versatile solutions in domains ranging from manufacturing to health care and beyond.

6.5 TRAINING AND OPTIMIZATION OF DEEP LEARNING MODELS

6.5.1 Data preparation and augmentation techniques for optical sensor data

The quality and diversity of training data are critical factors influencing the performance of machine learning models, especially when dealing with optical sensor data. Effective data preparation and augmentation techniques play a pivotal role in enhancing the robustness, generalization, and performance of models trained on optical sensor data. This section explores the importance of data preparation and augmentation, delves into various techniques, and provides references for further exploration.

Data preparation involves transforming raw data into a format suitable for model training. Augmentation, on the other hand, involves creating new variations of existing data by applying controlled transformations. Both processes are crucial for addressing challenges such as limited data availability, class imbalance, and enhancing model resilience to real-world variations.

Preprocessing techniques

1. **Normalization and standardization:** Rescaling features to a common range (normalization) or converting them to have zero mean and unit variance (standardization) helps models converge faster and prevents them from being dominated by features with larger scales.
2. **Feature engineering:** Extracting relevant features from raw optical sensor data can enhance model performance. Domain-specific knowledge plays a significant role in this process.

Data augmentation techniques

1. **Image rotation and flipping:** For image data captured by optical sensors, applying rotations and flips creates variations of the same scene, effectively increasing the diversity of the training dataset [21].
2. **Noise injection:** Adding random noise to the data helps the model become more robust to noisy real-world scenarios, commonly encountered in optical sensor data.
3. **Data cropping:** Cropping images to various sizes and orientations can simulate different perspectives and object scales, contributing to improved model generalization.
4. **Color and brightness adjustments:** Altering color balance, brightness, and contrast of images introduces diversity and makes the model more adaptable to varying lighting conditions.

Generative techniques

1. **GANs:** GANs can create synthetic data samples that closely resemble real optical sensor data. These generated samples can be used to augment the training dataset, enhancing the model's capacity to capture complex data distributions.
2. **Variational autoencoders (VAEs):** VAEs generate new data samples from the learned latent space, enabling the creation of diverse data points for training.

Data augmentation libraries

1. **Keras ImageDataGenerator:** A widely used library for data augmentation in image classification tasks.
2. **Albumentations:** A Python library for image and data augmentation, offering a wide range of transformations.

Data preparation and augmentation techniques are vital components in the success of machine learning models, especially when dealing with optical sensor data. Effective preprocessing methods and augmentation strategies ensure that models are trained on diverse, representative, and resilient datasets. These techniques enhance the model's ability to generalize to unseen scenarios, improve its performance, and make it robust in real-world applications.

6.5.2 Hyperparameter tuning and optimization strategies

The performance of machine learning models is heavily influenced by hyperparameters: parameters set before training that govern the learning process. Hyperparameter tuning and optimization are crucial steps to achieve the best possible model performance. The complex relationships between hyperparameters and model outcomes make efficient tuning essential. This section explores the significance of hyperparameter tuning, and common optimization strategies, and provides references for further exploration.

Hyperparameters control various aspects of model training, including regularization strength, learning rate, and architecture specifics. Selecting appropriate values significantly impacts model performance and generalization. Poorly chosen hyperparameters can lead to overfitting, slow convergence, or suboptimal results. Tuning hyperparameters requires a systematic approach to identify the best configuration for a given task.

- **Grid search:** Grid search involves specifying a grid of possible hyperparameter values and evaluating the model's performance for each combination. While systematic, it can be computationally expensive and may not scale well for large hyperparameter spaces.
- **Random search [36]:** Random search selects random combinations of hyperparameter values for evaluation. It is more efficient than grid search, as it allows for a broader exploration of the hyperparameter space.

Bayesian optimization [37] utilizes probabilistic models to estimate the performance landscape of different hyperparameter configurations. It intelligently selects the next configuration to evaluate, optimizing the model while minimizing the number of evaluations needed.

Gradient-based optimization methods utilize gradient information to update hyperparameters iteratively. Techniques like gradient descent and its variants aim to find optimal hyperparameters by minimizing an objective function.

Hyperparameter libraries

1. **Scikit-learn:** A popular Python library that offers grid search and randomized search capabilities for hyperparameter tuning.
2. **Hyperopt:** A library that employs Bayesian optimization for hyperparameter optimization.

Automated hyperparameter tuning

1. **AutoML:** Automated Machine Learning platforms, such as Google AutoML and Auto-Sklearn, provide end-to-end solutions that include hyperparameter tuning.
2. **Keras Tuner:** A hyperparameter tuning library for Keras that combines grid search and Bayesian optimization.

Hyperparameter tuning is a critical component of building high-performing machine learning models. The choice of hyperparameters significantly impacts the model's ability to generalize and perform well on unseen data. Strategies such as grid search, random search, Bayesian optimization, and gradient-based optimization offer a systematic approach to finding optimal hyperparameter configurations. With the availability of libraries and automated tools, hyperparameter tuning has become more accessible, enabling researchers and practitioners to extract the best possible performance from their models.

6.5.3 Transfer learning and pretrained models for optical sensor tasks

Transfer learning, a technique where knowledge gained from one task is leveraged to improve performance on a related task, has proven to be a powerful paradigm in the field of machine learning [29]. In recent years, pretrained models have emerged as a cornerstone of transfer learning, particularly in tasks involving optical sensor data. This section explores the significance of transfer learning and pretrained models for optical sensor tasks, discusses their applications, and provides references for further exploration.

Transfer learning addresses the challenge of limited labeled data by enabling models to learn from existing knowledge. Instead of starting from scratch, models benefit from features learned from large datasets and generalize this knowledge to perform well on new, related tasks. This approach is especially valuable in domains like optical sensor data analysis, where collecting labeled data can be resource-intensive.

Pretrained models, typically trained on massive datasets, have demonstrated remarkable capabilities in image understanding. These models have already learned complex features from diverse visual data and can be fine-tuned or adapted to specific optical sensor tasks, saving time and resources.

Applications in optical sensor tasks

1. **Image classification:** Pretrained models like VGG, ResNet, and Inception have shown exceptional accuracy in image classification tasks involving optical sensor data. By fine-tuning these models on specific classes, accuracy can be further improved.
2. **Object detection:** Faster R-CNN, YOLO, and Single Shot MultiBox Detector (SSD) architectures pretrained on extensive datasets can be

fine-tuned for object detection in optical sensor imagery. This is beneficial in applications like surveillance and autonomous vehicles.
3. **Semantic segmentation:** Pretrained models are invaluable for semantic segmentation tasks in which each pixel is classified. Models like U-Net [38], pretrained on diverse image datasets, can be adapted for accurate segmentation of objects in optical sensor data.

Fine-tuning and transfer learning strategies

1. **Feature extraction:** In this approach, the pretrained model's early layers, which capture low-level features, are retained, while later layers are replaced and fine-tuned for the specific task.
2. **Fine-tuning all layers:** All layers of the pretrained model can be fine-tuned on the target task using a smaller learning rate, updating weights to specialize the model.
3. **One-shot learning:** Pretrained models can be used with few-shot learning, where a minimal amount of labeled data is sufficient to adapt the model to the target task.

Transfer learning and pretrained models have revolutionized the field of optical sensor data analysis by enabling models to quickly adapt to specific tasks without requiring an abundance of labeled data. These models, with their ability to learn intricate features from large, diverse datasets, form a crucial foundation for accurate image classification, object detection, and semantic segmentation tasks. As research and advancements in pretrained models continue to evolve, they promise to play an even greater role in pushing the boundaries of what can be achieved with optical sensor data.

6.6 APPLICATIONS OF DEEP LEARNING IN OPTICAL SENSOR SYSTEMS

6.6.1 Deep learning-based object detection and recognition

Object detection and recognition are fundamental tasks in computer vision with wide-ranging applications, from autonomous vehicles to surveillance systems. Deep learning has revolutionized these tasks by enabling models to automatically learn relevant features from data, allowing for accurate and efficient detection and recognition of objects. This section explores the significance of deep learning-based object detection and recognition, discusses key concepts and techniques, and provides references for further exploration.

Traditional methods for object detection and recognition relied on handcrafted features and complex pipelines. Deep learning, particularly CNNs, has transformed these tasks by learning features directly from raw data. CNNs can capture both low-level visual patterns (edges, textures) and high-level semantic structures, leading to superior performance in object detection and recognition tasks.

Object detection involves identifying objects within images or videos and providing their spatial locations. Deep learning approaches have introduced remarkable accuracy and speed to this task. Some notable techniques include the following:

1. **Faster R-CNN** [39]: This architecture combines region proposal networks and CNNs to identify object locations and classes simultaneously.
2. **YOLO** (**you only look once**) [40]: YOLO models divide images into grids and predict object classes and bounding box coordinates in a single pass, making them incredibly fast for real-time applications.
3. **SSD**: SSD uses a set of predefined anchor boxes to predict object locations and classes, providing a balance between accuracy and speed.

Object recognition involves assigning a label to an object within an image. Deep learning-based recognition models excel due to their ability to capture complex visual patterns. Techniques include:

1. **CNN-based classification** [41]: CNN architectures, such as VGG, ResNet, and Inception, achieve state-of-the-art performance in image classification tasks.
2. **Transfer learning** [29]: Pretrained CNN models on large datasets can be fine-tuned for specific recognition tasks with limited labeled data, saving training time and improving accuracy.

While deep learning has achieved remarkable success, challenges persist, such as handling occlusions, scale variations, and small object detection. Advances like attention mechanisms, multiscale feature fusion, and improved architectures continue to address these challenges, pushing the boundaries of accuracy and robustness.

Deep learning-based object detection and recognition have transformed computer vision tasks, enabling accurate and efficient identification of objects in images and videos. Techniques like Faster R-CNN, YOLO, CNN-based classification, and transfer learning have revolutionized these tasks. As research in deep learning advances, object detection and recognition systems are poised to become even more precise, adaptive, and capable, fueling innovations across industries.

6.6.2 Deep learning for anomaly detection in optical sensor data

Anomaly detection is a critical task in various domains, including manufacturing, security, and health care, where identifying rare and unexpected events is crucial [26]. Deep learning has emerged as a powerful approach for anomaly detection in optical sensor data, enabling the automated discovery of irregular patterns that may indicate faults, intrusions, or anomalies. This section explores the significance of deep learning in anomaly detection for optical sensor data, discusses methodologies, and provides references for further exploration.

Traditional anomaly detection methods often rely on handcrafted features and statistical techniques that might struggle to capture complex patterns in high-dimensional data like optical sensor readings. Deep learning models, particularly autoencoders and RNNs, have proven effective in learning intricate representations and identifying anomalies.

Autoencoders are neural network architectures that learn to reconstruct input data from a reduced-dimensional representation called the latent space [42]. Anomalies can be detected when the reconstruction error is significantly higher for data points that deviate from the normal patterns learned during training.

RNNs are well-suited for time-series data, making them suitable for detecting anomalies in sequential optical sensor data. RNNs can capture temporal dependencies and patterns that may indicate anomalies, such as sudden spikes or drops in sensor readings.

Applications and advantages

1. **Manufacturing quality control**: In manufacturing, anomalies can indicate defects or faults in production processes. Deep learning models can identify anomalies in sensor data collected from manufacturing equipment, contributing to quality assurance.
2. **Cybersecurity**: Anomalies in network traffic, which can be detected through deep learning, could indicate potential security breaches or cyberattacks.
3. **Health-care monitoring**: Deep learning models applied to data from medical optical sensors can identify irregularities in vital signs, allowing for early detection of health issues.

Despite its effectiveness, deep learning for anomaly detection faces challenges, such as the need for large amounts of labeled data for training and the difficulty of determining appropriate thresholds for anomaly detection. Research is ongoing to develop models that are more robust, interpretable, and capable of handling varying levels of anomaly severity.

Deep learning has significantly enhanced anomaly detection in optical sensor data by automatically learning complex patterns and representations. Techniques like autoencoders and RNNs have proven valuable in diverse applications, offering advantages in manufacturing, cybersecurity, and health care. As research continues to advance, deep learning approaches for anomaly detection are likely to become even more accurate and adaptable, enabling industries to proactively identify and address anomalies to ensure safety, quality, and efficiency.

6.6.3 Deep learning for predictive maintenance and fault detection

Predictive maintenance and fault detection are essential practices in various industries to prevent equipment failures and optimize operational efficiency. Deep learning has emerged as a powerful tool for these tasks, enabling accurate predictions of machinery failures and early detection of faults. This section delves into the significance of deep learning in predictive maintenance and fault detection, explores methodologies and applications, and provides references for further exploration.

Traditional maintenance practices rely on scheduled maintenance routines, which can be inefficient and costly. Deep learning models can leverage historical sensor data to predict when equipment is likely to fail, allowing maintenance to be performed only when necessary. This approach reduces downtime and minimizes operational costs.

Deep learning models can identify patterns associated with faulty or abnormal equipment behavior. By learning from labeled data, these models can differentiate between normal operating conditions and deviations that may indicate faults or impending failures.

Applications and approaches

1. **Predictive maintenance in manufacturing** [43]: Deep learning models can analyze sensor data from manufacturing equipment to predict failures, enabling timely maintenance and reducing production interruptions.
2. **Aircraft engine health monitoring** [44]: Deep learning can analyze sensor data from aircraft engines to predict potential malfunctions, ensuring passenger safety and operational efficiency.
3. **Energy grid fault detection:** Deep learning models can analyze data from smart grids to detect anomalies, helping utilities identify and address potential faults or irregularities.

Deep learning for predictive maintenance and fault detection comes with challenges, including the need for labeled data, model interpretability, and handling imbalanced datasets. It's essential to strike a balance between model complexity and interpretability, especially in safety-critical applications.

As deep learning techniques continue to evolve, the integration of domain knowledge and explainable artificial intelligence (AI) will be crucial. Hybrid approaches that combine physics-based models with data-driven deep learning methods can lead to more accurate and reliable predictions and fault detection.

Deep learning has revolutionized predictive maintenance and fault detection by enabling accurate predictions and early detection of equipment failures. By learning intricate patterns from historical sensor data, deep learning models can help industries avoid costly downtime, optimize maintenance schedules, and ensure the safety and efficiency of operations. As research advances and challenges are addressed, deep learning's role in predictive maintenance and fault detection is set to become even more significant, benefiting various industries worldwide.

6.7 CHALLENGES AND FUTURE DIRECTIONS

Deep learning has revolutionized various fields, including optical sensor applications, by enabling automated and accurate data analysis and interpretation [45]. However, despite its remarkable success, deep learning is not without its challenges and limitations. This section discusses some of the key challenges and limitations associated with the application of deep learning in optical sensor tasks, exploring issues such as data limitations, interpretability, computational demands, and overfitting, and providing references for further exploration.

Deep learning models thrive on large amounts of labeled data. In many optical sensor applications, obtaining sufficient and diverse labeled data can be challenging and expensive. Furthermore, the quality and accuracy of labels are crucial, as incorrect labels can lead to biased or erroneous models.

Deep learning models are often considered black boxes due to their complex architectures. Understanding how and why a model makes certain predictions is crucial, especially in applications where decisions have significant consequences. Interpretable and explainable AI techniques are necessary to build trust and gain insights from the model's predictions.

Deep learning models are often computationally intensive, requiring significant processing power and memory. Deploying such models on resource-constrained devices or in real-time applications can be challenging due to latency and energy consumption concerns.

Deep learning models are prone to overfitting, where they learn noise and irrelevant patterns from the training data, leading to poor generalization on unseen data. Regularization techniques and carefully curated datasets are essential to mitigate this challenge.

Deep learning models are vulnerable to adversarial attacks, where small, carefully crafted perturbations to input data can cause the model to make incorrect predictions [46]. Robustness against such attacks is a critical concern, especially in security-sensitive applications.

Deep learning models trained on one domain may not perform well on a different domain. Transfer learning can mitigate this, but finding the right balance between pretraining and fine-tuning can be challenging.

While deep learning has shown exceptional promise in optical sensor applications, its challenges and limitations must be acknowledged. These limitations span data availability, interpretability, computational requirements, generalization, adversarial attacks, and domain specificity. Overcoming these challenges demands a holistic approach that integrates domain expertise, careful model design, novel regularization techniques, and a strong emphasis on ethical and explainable AI. As researchers and practitioners work to address these limitations, the potential of deep learning in optical sensor applications remains immense, offering innovative solutions to complex problems.

6.7.1 Emerging trends and advancements in the field

The landscape of technological innovation is constantly evolving, driven by ongoing research, development, and the convergence of various disciplines [47]. Emerging trends and advancements in diverse fields are shaping the way we live, work, and interact. This section delves into some of the most notable emerging trends and advancements across different domains, providing insights into their impact and references for further exploration.

AI and machine learning have witnessed significant advancements, with models like GPT-3, BERT, and neural architecture search (NAS) leading the way [48]. These models are revolutionizing natural language processing, understanding, and generation, reshaping the landscape of communication, automation, and decision-making.

Quantum computing is rapidly advancing, offering the potential to solve complex problems currently beyond classical computers' capabilities [49]. Quantum supremacy milestones and advancements in quantum error correction bring us closer to practical quantum computing applications in cryptography, optimization, and material science.

Advancements in biotechnology, including CRISPR-Cas9 gene editing and synthetic biology, are revolutionizing health care, agriculture, and environmental conservation [50]. These technologies hold the promise of personalized medicine, disease eradication, and sustainable food production [51].

Advancements in renewable energy technologies, such as solar cells, wind turbines, and energy storage solutions, are driving the transition toward cleaner and more sustainable energy sources [52]. Innovations in materials, efficiency, and grid integration are making renewable energy more accessible and cost-effective [53].

Emerging trends and advancements in various fields are reshaping our world in profound ways. AI, quantum computing, biotechnology, and renewable energy are just a few examples of the transformative technologies that are paving the way for a more interconnected, efficient, and sustainable future. These trends, driven by ongoing research and innovation, are set to impact every aspect of our lives, from communication and health care to energy and beyond.

6.7.2 Future research directions and potential innovations

The ever-evolving landscape of science and technology continually opens doors to new research directions and potential innovations [54]. As our understanding deepens and capabilities expand, exciting possibilities emerge across various fields. This section delves into the future research directions and potential innovations that hold the promise of shaping our world in the coming years, providing insights into key areas and references for further exploration.

Future research in AI and machine learning is likely to focus on developing more robust, explainable, and ethical AI systems [55]. This involves addressing bias and fairness concerns, creating models that can reason and understand context, and enhancing the interpretability of complex neural networks.

Future innovations in biotechnology may include advancements in personalized medicine, such as gene therapies and precision diagnostics, as well as breakthroughs in synthetic biology, enabling the creation of novel organisms with specific functions for applications in medicine, agriculture, and environmental remediation [56].

Quantum computing research will likely focus on error correction, building larger and more stable qubit systems, and exploring new quantum algorithms for optimization, cryptography, and material simulations [57]. Practical applications of quantum computing in chemistry and drug discovery are also anticipated.

Future research in sustainability will involve innovations in renewable energy technologies, energy storage solutions, carbon capture and utilization, and sustainable materials [58]. Advancements in climate modeling and policy development are crucial for mitigating the impacts of climate change.

In space exploration, research will focus on developing technologies for sustainable lunar and Martian habitats, resource extraction from asteroids, and advanced propulsion systems [59]. These innovations are critical for humanity's expansion beyond Earth.

The future of research and innovation holds boundless potential across various domains. From enhancing AI to tackling climate change and exploring the cosmos, researchers and scientists are poised to drive advancements that will shape our world for generations to come. By focusing on ethics, sustainability, and collaboration, these future research directions and potential innovations have the power to solve complex challenges and create a more prosperous and interconnected global society.

6.8 CASE STUDIES AND PRACTICAL IMPLEMENTATIONS

Deep learning has revolutionized the capabilities of optical sensor systems, enabling them to extract intricate information from visual data and make informed decisions in real time. From autonomous vehicles to medical imaging, deep learning has found applications in various domains. This section explores real-world examples of deep learning–enabled optical sensor systems, highlighting their impact and providing references for further exploration.

Deep learning plays a crucial role in enabling the perception and decision-making capabilities of autonomous vehicles [60]. Cameras, LiDAR, and other optical sensors capture real-time data, which deep learning models process to recognize objects, pedestrians, road signs, and even predict the behavior of other vehicles, ensuring safe navigation.

Deep learning has transformed medical imaging by enhancing diagnosis accuracy and efficiency [61]. Optical sensors in devices like MRIs and CT scanners capture detailed images, which deep learning models analyze to detect abnormalities, segment organs, and assist radiologists in making precise diagnoses.

In manufacturing, optical sensor systems integrated with deep learning models can detect defects in products with high accuracy [62]. Cameras capture images of items on production lines, and deep learning algorithms analyze the images to identify imperfections, ensuring quality control.

Deep learning–enabled optical sensors aid in monitoring crops and assessing their health [63]. Drones equipped with cameras capture images of fields, which deep learning models analyze to detect disease, predict yields, and optimize irrigation and fertilizer usage.

Deep learning-enhanced optical sensors are deployed for environmental monitoring, such as detecting pollution levels and monitoring wildlife [64]. Cameras and satellite imagery are processed by deep learning models to track changes in ecosystems, contributing to conservation efforts.

Deep learning–enabled optical sensor systems have brought transformative capabilities to a wide range of applications. From autonomous vehicles ensuring safe transportation to medical imaging enhancing health-care diagnostics, these systems leverage deep learning's ability to extract meaningful insights from visual data. As technology continues to advance, these real-world examples demonstrate the potential for deep learning to address complex challenges and revolutionize diverse industries.

6.8.1 Successful applications and case studies in various industries

Innovations driven by technology have paved the way for successful applications and case studies across diverse industries. From health care to finance, these real-world examples highlight how cutting-edge solutions are transforming processes, improving efficiency, and enhancing user experiences. This section delves into successful applications and case studies in various industries, offering insights into their impact and providing references for further exploration.

Advancements in medical imaging and diagnostics have revolutionized health care [65]. For instance, IBM's Watson for Oncology utilizes natural language processing and machine learning to assist oncologists in diagnosing and recommending personalized treatment plans for cancer patients.

Financial institutions deploy sophisticated algorithms for fraud detection. Mastercard's Decision Intelligence [66] system employs machine learning to assess billions of transactions in real time, flagging potential fraud while minimizing false positives and improving customer experience.

Predictive maintenance is transforming manufacturing operations. Siemens' MindSphere [67] platform uses Internet of Things (IoT) data and machine learning to predict equipment failures, optimizing maintenance schedules and reducing downtime.

Precision farming leverages technology to enhance agricultural practices. John Deere's See & Spray [68] technology employs computer vision and machine learning to identify and target individual plants, reducing herbicide use and promoting sustainable farming.

Smart grids integrate technology for efficient energy management. Pacific Gas and Electric Company (PG&E) [69] utilizes advanced metering infrastructure and data analytics to optimize energy distribution, reduce outages, and enhance grid reliability.

Ride-sharing services have transformed transportation. Uber's [70] platform employs data analytics and machine learning to optimize ride allocation, reduce wait times, and enhance overall user experience.

Successful applications and case studies across industries highlight the transformative power of technology-driven solutions. From health care and finance to manufacturing and agriculture, these examples demonstrate how innovation is driving efficiency, accuracy, and sustainability. As technology continues to evolve, these successes serve as a testament to the potential for creating positive change across various sectors, improving processes, and enhancing quality of life for people around the world.

6.9 CONCLUSION AND KEY TAKEAWAYS

In this chapter, we explored the transformative impact of deep learning in various fields, focusing on its applications in optical sensor systems. We discussed how deep learning has revolutionized object detection and recognition, anomaly detection, predictive maintenance, and fault detection in optical sensor data.

The chapter began by highlighting the significance of deep learning in optical computing, emphasizing its ability to handle complex data and improve the efficiency of optical systems. We then delved into different deep learning architectures relevant to optical sensor data, including CNNs for image-based sensors and RNNs for time-series data.

We explored other deep learning architectures beyond CNNs and RNNs, including GANs for data generation, attention mechanisms for enhanced feature extraction, and transformer models for sequence tasks. These architectures contribute to more accurate and versatile analysis of optical sensor data.

A critical aspect discussed was the importance of data preparation and augmentation techniques in enhancing deep learning model performance. We covered strategies such as data normalization, augmentation, and handling imbalanced datasets, all of which are vital for training robust models.

Hyperparameter tuning and optimization strategies were addressed as essential components of deep learning model development. Techniques like grid search, random search, and Bayesian optimization were explored, emphasizing their role in finding optimal model configurations.

Transfer learning and pretrained models emerged as powerful tools for accelerating model development in optical sensor tasks. We discussed how pretrained models can be fine-tuned for specific tasks, saving time and resources while achieving high performance.

The chapter also highlighted the challenges and limitations of deep learning in optical sensor applications, including issues related to data availability, interpretability, computational demands, overfitting, and adversarial attacks. These challenges underscored the need for careful consideration and ongoing research.

Lastly, we explored emerging trends, future research directions, and potential innovations. From AI and quantum computing to biotechnology and sustainability, the future promises groundbreaking advancements that will shape industries and societies.

Overall, the chapter showcased the diverse applications and promising potential of deep learning in optical sensor systems, underscoring its role in driving innovation and addressing complex challenges in various domains.

6.9.1 Key takeaways and implications for the future

The exploration of deep learning in optical sensor applications has revealed several key takeaways and important implications for the future:

1. **Enhanced data analysis:** Deep learning has proven to be a game-changer in the analysis of optical sensor data. Its ability to automatically extract intricate patterns from complex data has opened avenues for accurate and efficient decision-making.
2. **Accurate detection and recognition:** Deep learning, particularly CNNs, has enabled highly accurate object detection and recognition in images obtained from optical sensors. This has implications across industries, from manufacturing to health care, where precise identification of objects is crucial.
3. **Proactive maintenance:** The utilization of deep learning for predictive maintenance and fault detection in optical sensor systems is reducing downtime and operational costs across sectors. By identifying

anomalies and predicting failures, organizations can optimize maintenance schedules and avoid unexpected breakdowns.

4. **Real-time anomaly detection:** Deep learning–powered anomaly detection is enabling real-time monitoring of optical sensor data. This is especially valuable in critical sectors like cybersecurity, where rapid response to anomalies is vital for preventing potential threats.
5. **Transfer learning's efficiency:** The adoption of transfer learning and pretrained models is accelerating model development in optical sensor tasks. By leveraging preexisting knowledge from large datasets, developers can fine-tune models for specific applications, saving time and resources.
6. **Interdisciplinary collaboration:** The success of deep learning in optical sensor applications underscores the importance of interdisciplinary collaboration. Engineers, data scientists, domain experts, and ethicists must collaborate to ensure the effectiveness, safety, and ethical use of these technologies.
7. **Ethical considerations:** As deep learning's influence grows, ethical considerations become more crucial. Ensuring fair and unbiased model behavior, addressing privacy concerns, and creating transparent and interpretable models are imperative.
8. **Future innovations:** Emerging trends such as quantum computing, AI explainability, and sustainability will drive future innovations in optical sensor applications. Researchers and practitioners must stay updated on these trends to harness their full potential.
9. **Data availability:** Access to high-quality labeled data remains a challenge in deep learning. Industries should invest in data collection and annotation processes to enable the development of accurate and reliable models.
10. **Robustness and security:** Developing deep learning models that are robust against adversarial attacks and capable of handling uncertainties will be a key focus for future research. Ensuring the security and reliability of deep learning–enabled systems is paramount.

In the future, deep learning's role in optical sensor applications will likely continue to expand, leading to safer and more efficient technologies. As we navigate this technological landscape, addressing challenges, fostering collaboration, and upholding ethical considerations will be pivotal in harnessing the full potential of deep learning in optical sensor systems.

REFERENCES

[1] Fournier, Maryse, Pierre Barritault, Gabriel Jobert, Adrien Marchant, Salim Boutami, Julien Michelot, Pierre Lienhard, Sergio Nicoletti, and Laurent Duraffourg. "A miniaturized optical sensor for particulate matter detection."

In *Photonic Instrumentation Engineering VII*, vol. 11287, pp. 261–268. SPIE, 2020.

[2] Gao, Xiao-Guang, Ling-Xiao Cheng, Wen Shuai Jiang, Xiao-Kuan Li, and Fei Xing. "Graphene and its derivatives-based optical sensors." *Frontiers in Chemistry* 9 (2021): 615164.

[3] Kim, Michele M., and Arash Darafsheh. "Light sources and dosimetry techniques for photodynamic therapy." *Photochemistry and Photobiology* 96, no. 2 (2020): 280–294.

[4] Wu, Bolang, Ge Zhu, Alain Dufresne, and Ning Lin. "Fluorescent aerogels based on chemical crosslinking between nanocellulose and carbon dots for optical sensor." *ACS Applied Materials & Interfaces* 11, no. 17 (2019): 16048–16058.

[5] Krutz, David, Rupert Müller, Uwe Knodt, Burghardt Günther, Ingo Walter, Ilse Sebastian, Thomas Säuberlich et al. "The instrument design of the DLR earth sensing imaging spectrometer (DESIS)." *Sensors* 19, no. 7 (2019): 1622.

[6] Tian, Jingyi, Qiang Li, Pavel A. Belov, Ravindra Kumar Sinha, Weiping Qian, and Min Qiu. "High-Q all-dielectric metasurface: super and suppressed optical absorption." *Acs Photonics* 7, no. 6 (2020): 1436–1443.

[7] Liu, Tingting, Hai Liu, You-Fu Li, Zengzhao Chen, Zhaoli Zhang, and Sannyuya Liu. "Flexible FTIR spectral imaging enhancement for industrial robot infrared vision sensing." *IEEE Transactions on Industrial Informatics* 16, no. 1 (2019): 544–554.

[8] Roriz, Paulo, Susana Silva, Orlando Frazão, and Susana Novais. "Optical fiber temperature sensors and their biomedical applications." *Sensors* 20, no. 7 (2020): 2113.

[9] Rodriguez-Saona, Luis, Didem Peren Aykas, Karla Rodrigues Borba, and Alejandra Urtubia. "Miniaturization of optical sensors and their potential for high-throughput screening of foods." *Current Opinion in Food Science* 31 (2020): 136–150.

[10] Kot, Patryk, Magomed Muradov, Michaela Gkantou, George S. Kamaris, Khalid Hashim, and David Yeboah. "Recent advancements in non-destructive testing techniques for structural health monitoring." *Applied Sciences* 11, no. 6 (2021): 2750.

[11] Hong, Danfeng, Wei He, Naoto Yokoya, Jing Yao, Lianru Gao, Liangpei Zhang, Jocelyn Chanussot, and Xiaoxiang Zhu. "Interpretable hyperspectral artificial intelligence: When nonconvex modeling meets hyperspectral remote sensing." *IEEE Geoscience and Remote Sensing Magazine* 9, no. 2 (2021): 52–87.

[12] Chen, Zailiang, Ziyang Zeng, Hailan Shen, Xianxian Zheng, Peishan Dai, and Pingbo Ouyang. "DN-GAN: Denoising generative adversarial networks for speckle noise reduction in optical coherence tomography images." *Biomedical Signal Processing and Control* 55 (2020): 101632.

[13] Ma, Jiyang, Jianming Wen, Shulin Ding, Shengjun Li, Yong Hu, Xiaoshun Jiang, Liang Jiang, and Min Xiao. "Chip-Based Optical Isolator and Nonreciprocal Parity-Time Symmetry Induced by Stimulated Brillouin Scattering." *Laser & Photonics Reviews* 14, no. 5 (2020): 1900278.

[14] Xu, Jun, Ting Li, Yitong Li, Cheng Zhang, Ling Cheng, Li Liu, and Changyun Miao. "Fabrication and application of a grooved optical fiber respiratory sensor based on geometric parameter optimization by optical simulation." *Optics & Laser Technology* 167 (2023): 109747.

[15] Thurner, Klaus, Pierre-François Braun, and Khaled Karrai. "Fabry-Pérot interferometry for long range displacement sensing." *Review of Scientific Instruments* 84, no. 9 (2013).
[16] Kadhim, Riadh A., Abdul Kareem K. and Liming Yuan. "Advances in surface plasmon resonance-based plastic optical fiber sensors." *IETE Technical Review* 39, no. 2 (2022): 442–459.
[17] Chakraborty, Shubhankar, and Das P. K. "Characterisation and classification of gas-liquid two-phase flow using conductivity probe and multiple optical sensors." *International Journal of Multiphase Flow* 124 (2020): 103193.
[18] Chu, Yuan, Feng Xu, Tianhuai Ding, and Shujian Chen. "Measurement of rotation axis transverse displacements based on diffuse reflective intensity modulated optical fiber sensor." *Optical Engineering* 45, no. 8 (2006): 084403.
[19] Zhao, Zhiqiang, W. S. Lau, Albert CK Choi, and Y. Y. Shan. "Modulation functions of the reflective optical fiber sensor for specular and diffuse reflection." *Optical Engineering* 33, no. 9 (1994): 2986–2991.
[20] Patil, Supriya S., and A. D. Shaligram. "Modeling and experimental studies on retro-reflective fiber optic micro-displacement sensor with variable geometrical properties." *Sensors and Actuators A: Physical* 172, no. 2 (2011): 428–433.
[21] Krizhevsky, A., Sutskever, I., and Hinton, G. E. "ImageNet classification with deep convolutional neural networks." *Advances in Neural Information Processing Systems*, 25 (2012).
[22] Hochreiter, S., and Schmidhuber, J. "Long short-term memory." *Neural Computation* 9, no. 8 (2014): 1735–1780.
[23] Goodfellow, I., Pouget-Abadie, J., Mirza, M., Xu, B., Warde-Farley, D., Ozair, S., ... and Bengio, Y. "Generative adversarial nets." *Advances in Neural Information Processing Systems* 27 (2014).
[24] Vaswani, A., Shazeer, N., Parmar, N., Uszkoreit, J., Jones, L., Gomez, A. N., ... and Polosukhin, I. "Attention is all you need." *Advances in Neural Information Processing Systems* 30 (2017).
[25] Yosinski, J., Clune, J., Bengio, Y., and Lipson, H. "How transferable are features in deep neural networks?" *Advances in Neural Information Processing Systems* 27 (2014).
[26] Hinton, G. E., and Salakhutdinov, R. R. "Reducing the dimensionality of data with neural networks." *Science* 313, no. 5786 (2006): 504–507.
[27] Simonyan, K., and Zisserman, A. "Very deep convolutional networks for large-scale image recognition." arXiv preprint arXiv:1409.1556 v.
[28] Szegedy, C., Vanhoucke, V., Ioffe, S., Shlens, J., and Wojna, Z. "Rethinking the inception architecture for computer vision." In *Proceedings of the IEEE conference on computer vision and pattern recognition* (pp. 2818–2826) (2016).
[29] He, K., Zhang, X., Ren, S., and Sun, J. "Deep residual learning for image recognition." In *Proceedings of the IEEE conference on computer vision and pattern recognition* (pp. 770–778) (2016).
[30] Lipton, Z. C., Berkowitz, J., and Elkan, C. "A critical review of recurrent neural networks for sequence learning." arXiv preprint arXiv:1506.00019. (2015).
[31] Greff, K., Srivastava, R. K., Koutník, J., Steunebrink, B. R., & Schmidhuber, J. "LSTM: A search space odyssey." *IEEE Transactions on Neural Networks and Learning Systems* 28, no. 10 (2017): 2222–2232.
[32] Chandra, S., and Pant, M. "Recurrent neural networks (RNNs) for time series forecasting: Current status and future directions." *Expert Systems with Applications* 140 (2020): 112886.

[33] Cho, K., Van Merriënboer, B., Gulcehre, C., Bougares, F., Schwenk, H., and Bengio, Y. "Learning phrase representations using RNN encoder–decoder for statistical machine translation." arXiv preprint arXiv:1406.1078. (2014).
[34] Sabour, S., Frosst, N., and Hinton, G. E. "Dynamic routing between capsules." In *Advances in neural information processing systems* (pp. 3859–3869) (2017).
[35] Kingma, D. P., and Welling, M. "Auto-encoding variational bayes." arXiv preprint arXiv:1312.6114. (2013).
[36] Bergstra, J., and Bengio, Y. "Random search for hyper-parameter optimization." *Journal of Machine Learning Research*, 13(Feb) (2012):281–305.
[37] Snoek, J., Larochelle, H., and Adams, R. P. "Practical Bayesian optimization of machine learning algorithms." In *Advances in neural information processing systems* (pp. 2951–2959), (2012).
[38] Ronneberger, O., Fischer, P., and Brox, T. "U-Net: Convolutional networks for biomedical image segmentation." In *International Conference on Medical Image Computing and Computer-Assisted Intervention* (pp. 234–241), (2015).
[39] Ren, S., He, K., Girshick, R., and Sun, J. "Faster R-CNN: Towards real-time object detection with region proposal networks." In *Advances in neural information processing systems* (pp. 91–99), (2015).
[40] Redmon, J., Divvala, S., Girshick, R., and Farhadi, A. "You only look once: Unified, real-time object detection." In *Proceedings of the IEEE conference on computer vision and pattern recognition* (pp. 779–788), (2016).
[41] Simonyan, K., and Zisserman, A. "Very deep convolutional networks for large-scale image recognition." arXiv preprint arXiv:1409.1556 (2014).
[42] Malhotra, P., Ramakrishnan, A., Anand, G., Vig, L., Agarwal, P., and Shroff, G. "LSTM-based encoder-decoder for multi-sensor anomaly detection." In *Proceedings of the 2016 SIAM International Conference on Data Mining* (pp. 379–387), (2016).
[43] Yoon, J., and Yoon, J. "Predictive maintenance: A focused review." *Journal of Manufacturing Science and Engineering* 140, no. 5 (2018): 051004.
[44] Li, Z., Gao, R. X., and Wang, X. "A review of machine learning methods for the condition monitoring of mechanical systems." *Mechanical Systems and Signal Processing* 104 (2017): 799–834.
[45] Caruana, R., Lou, Y., Gehrke, J., Gehrke, J., Koch, P., Sturm, M., and Elhadad, N. "Intelligible models for healthcare: Predicting pneumonia risk and hospital 30-day readmission." *Proceedings of the 21th ACM SIGKDD International Conference on Knowledge Discovery and Data Mining*, (pp. 1721–1730) (2015).
[46] Szegedy, C., Zaremba, W., Sutskever, I., Bruna, J., Erhan, D., Goodfellow, I., and Fergus, R. "Intriguing properties of neural networks." arXiv preprint arXiv:1312.6199 (2013).
[47] Brown, T. B., Mann, B., Ryder, N., Subbiah, M., Kaplan, J., Dhariwal, P., ... and Amodei, D. "Language models are few-shot learners." arXiv preprint arXiv:2005.14165 (2020).
[48] Devlin, J., Chang, M. W., Lee, K., and Toutanova, K. "BERT: Pre-training of deep bidirectional transformers for language understanding." arXiv preprint arXiv:1810.04805 (2018).
[49] Arute, F., Arya, K., Babbush, R., Bacon, D., Bardin, J. C., Barends, R., ... and Chen, Y. "Quantum supremacy using a programmable superconducting processor." *Nature* 574, no. 7779 (2019), 505–510.

[50] Jinek, M., Chylinski, K., Fonfara, I., Hauer, M., Doudna, J. A., and Charpentier, E. "A programmable dual-RNA–guided DNA endonuclease in adaptive bacterial immunity." *Science* 337, no. 6096 (2012): 816–821.
[51] Purnick, P. E., and Weiss, R. "The second wave of synthetic biology: from modules to systems." *Nature Reviews Molecular Cell Biology* 10, no. 6 (2009): 410–422.
[52] Green, M. A., Ho-Baillie, A., and Snaith, H. J. "The emergence of perovskite solar cells." *Nature Photonics* 8, no. 7 (2014): 506–514.
[53] Zhang, X., and Zhao, X. "Recent advances in organic materials for organic solar cells." *Materials Today* 27 (2019): 36–55.
[54] Mitchell, M., Wu, S., Zaldivar, A., Barnes, P., Vasserman, L., Hutchinson, B., ... & Daumeé III, H. Model cards for model reporting. In *Proceedings of the conference on fairness, accountability, and transparency* (pp. 220–229), 2019.
[55] Lake, B. M., Ullman, T. D., Tenenbaum, J. B., & Gershman, S. J. "Building machines that learn and think like people." *Behavioral and Brain Sciences* 40 (2017): e253.
[56] Wu, Y., Zeng, J., Rosser, S. J., and Yang, M. "Synthetic biology: from the digital to the natural." *Frontiers in Bioengineering and Biotechnology* 9 (2021): 668597.
[57] Preskill, J. (2018). Quantum computing in the NISQ era and beyond. Quantum, 2, 79.
[58] IEA. (2021). Net Zero by 2050: A Roadmap for the Global Energy Sector.
[59] Zubrin, Robert, Richard Wagner, and Christopher P. McKay. "The case for Mars: the plan to settle the red planet and why we must." *Nature* 383, no. 6603 (1996): 780–780.
[60] Bojarski, Mariusz, et al. "End to end learning for self-driving cars." arXiv preprint arXiv:1604.07316 (2016).
[61] Esteva, Andre, et al. "Dermatologist-level classification of skin cancer with deep neural networks." *Nature* 542, no. 7639 (2017): 115–118.
[62] Oh, Dong Yul, and Il Dong Yun. "Residual error based anomaly detection using auto-encoder in SMD machine sound." *Sensors* 18, no. 5 (2018): 1308.
[63] Abeyrathna, R.M.D. et al. "Recognition and counting of apples in a dynamic state using a 3D camera and deep learning algorithms for robotic harvesting systems." *Sensors* 23, no. 8 (2023): 3810.
[64] Norouzzadeh, Mohammad Sadegh, et al. "Automatically identifying, counting, and describing wild animals in camera-trap images with deep learning." *Proceedings of the National Academy of Sciences* 115, no. 25 (2018): E5716–E5725.
[65] Tricco, Andrea C., et al. "A scoping review on the conduct and reporting of scoping reviews." *BMC Medical Research Methodology* 16 (2016): 1–10.
[66] Mastercard. (2021). "Mastercard announces Decision Intelligence." [Online] Available at: https://www.mastercard.us/en-us/about-mastercard/what-we-do/decision-intelligence.html
[67] Siemens. (2021). "Siemens' MindSphere Industrial IoT as a service on Azure." [Online] Available at: https://azure.microsoft.com/en-us/services/iot-edge/iot-edge-on-windows/
[68] John Deere. (2021). "See & Spray Select." [Online] Available at: https://www.deere.com/en/technology-products/precision-agriculture/precision-application/see-and-spray/

[69] Pacific Gas and Electric Company. (2021). "SmartMeter™ Benefits." [Online] Available at: https://www.pge.com/en_US/small-medium-business/your-account/your-bill/understand-your-bill/smartmeter-benefits.page
[70] Uber. (2021). "Using AI to create efficient routes and reduce pickup times." [Online] Available at: https://www.uber.com/newsroom/using-ai-to-create-efficient-routes-and-reduce-pickup-times/

Chapter 7

An introduction to neuromorphic computing

Kamal Kishor Choure, Ankur Saharia, Nitesh Mudgal, Manisha Prajapat, Manish Tiwari and Ghanshyam Singh

7.1 INTRODUCTION

Neuromorphic computing belongs to the development of technology that copies the structures found in the brain to create models of neural systems. Over time, this concept expanded to include computing systems that can carry out computational models influenced by biology, such as neural networks and deep learning networks. Neuromorphic computing seeks to integrate key characteristics of natural neural structures into a new computer model, which challenges the traditional von Neumann model. This includes things like combining memory and processing, complex parallelism, large amounts of feedback, widespread network connectivity, low precision, random computation, and constant adaptation that looks like learning processes. The technique also includes sparse, spike-based interactions for distributed communication. Spiking neural networks (SNNs) use less energy when they are not working because they go into inactive states and work quickly because they work in asynchronous, event-driven ways [1]. Ongoing research explores emerging memory devices as potential synaptic components in neuromorphic computing. The hardware for neuromorphic computing demands attributes such as large memory storage, high speed, operating at low power, and a distinctive feature that replicates the functioning of the human brain. Memristor devices support neuromorphic computing due to their favorable characteristics aligning with these requirements [2]. While neuromorphic computers have traditionally focused on applications within machine learning and neuroscience, commonly referred to as "cognitive" applications, their computational features make them appealing for a broad range of computational problems [3]. Deep learning (DL) has significantly accelerated to the promotion of numerous techniques in present time. However, the exceptional efficiency of this delivery comes at a significant computational price, which presents a significant challenge in the process of adapting DL applications to common hardware platforms [4]. Advancements in neurological science have made great strides in recent years, continuously uncovering novel attributes of biological neurons. Recent advancements in technology allow for the precise capture of

DOI: 10.1201/9781003591030-7

temporal changes in the brain's internal activity, aiding in the understanding of how brain activity relates to the perception of certain stimuli [5]. A spiking neural method is a novel computational model of an artificial neural network that encodes and analyzes neural input using precisely timed spike trains [6]. In order to create real-world, completely autonomous substances, it is necessary to design algorithms that can generalize from experience and be applicable to a wide range of machine learning tasks, such as those in supervised and reinforcement learning. Parametric function approximators have the ability to provide such capabilities, as they can either simulate the environment or the agent's policy [7]. Although neuromorphic systems possess energy-efficient properties, they also present their own unique obstacles. A major obstacle in the field of neuromorphic computing is establishing relations of the complex neuromorphic algorithms [8].

7.2 ASSOCIATED TECHNIQUES OF NEUROMORPHIC COMPUTING

7.2.1 Convolutional neural networks

The primary objective of neuromorphic systems is to optimize the efficiency of information processing and minimize power consumption, while simultaneously increasing computing capabilities. Convolutional neural networks (CNNs) have significantly transformed neuromorphic computing by excelling at complex tasks such as image and pattern recognition. CNNs, which are specifically designed for the analysis of organized grid-like data like photographs, are a subset of DL models. The general schematic of CNN is depicted in Figure 7.1. CNNs use convolution operations to extract spatial characteristics using adjustable filters, which are interconnected layers such as convolutional, pooling, and fully connected layers. Pooling layers are used to rationalize the dimensions of feature maps, whereas fully linked layers are employed to do classification or regression tasks using the collected features. The integration of CNNs into neuromorphic computer architecture enables the replication of the parallelism and flexibility that are inherent in

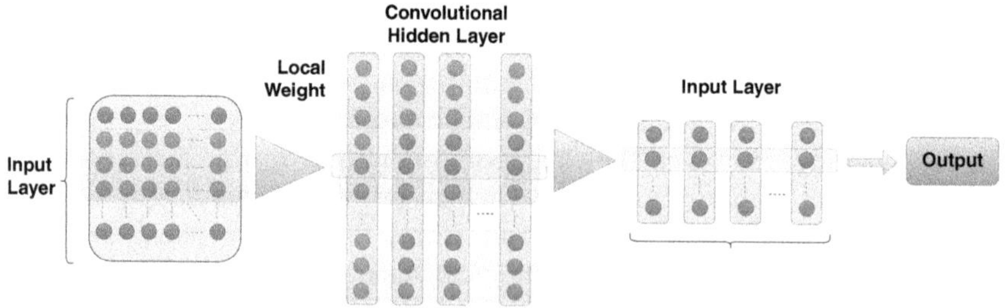

Figure 7.1 General schematic of CNN.

biological neural networks. This integration enables complex calculations with minimum energy use, making them especially well-suited for edge computing and AI applications in resource-constrained contexts. Moreover, focused research efforts are aimed at developing specialized neuromorphic hardware accelerators that are specifically tuned for CNNs [9, 10].

7.2.2 Spiking neural networks

The use of SNNs signifies a notable development within the realm of neuromorphic computing. In contrast to its predecessors, namely artificial neural networks (ANNs) and CNNs, which depend on continuous activation functions, SNNs function based on a concept derived from the firing patterns seen in real neurons, namely spikes or action potentials. SNNs may closely mimic the brain's asynchronous and event-driven processing via spike-based communication. SNNs use ANNs to mimic biological neurons by producing spikes in response to input inputs. The schematic of SNN is shown in Figure 7.2. The accumulation of receiving spikes by each neuron occurs gradually, and upon reaching a certain threshold, the neuron fires an output spike, therefore transmitting information to its interconnected neurons. Spike timing dependent plasticity (STDP) is a key principle in synapse neural networks that governs the alteration of synaptic weights in response to the relative timing of pre- and postsynaptic events. SNNs have the capability to acquire knowledge and adjust to input patterns, similar to the brain's

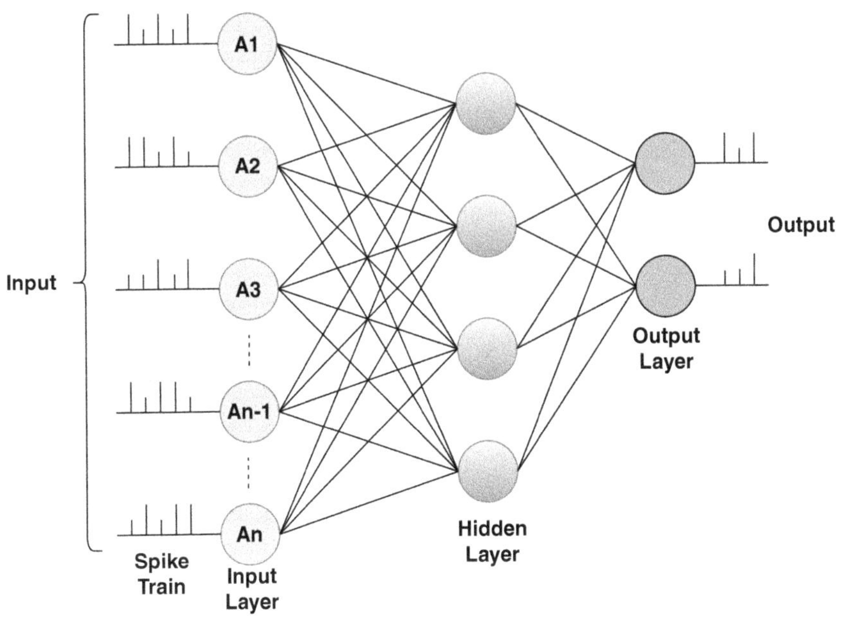

Figure 7.2 General schematic of SNN.

capacity for synaptic plasticity. An exceptional benefit of SNNs is their unmatched energy efficiency. SNNs have lower power consumption than conventional computer architecture due to their event-driven operation. This makes them well-suited for battery-powered devices and edge computing applications [11–13].

7.2.3 Reservoir computing

Reservoir computing is a novel computational paradigm that draws inspiration from the unique efficiency and parallel processing capabilities seen in the human brain. The reservoir is a dynamic system consisting of linked nodes or units, which is central to this paradigm. The nodes included in this system participate in complicated interactions, resulting in sophisticated temporal dynamics that are well-suited for computing tasks. Reservoir computing differs from traditional neural networks in that it does not modify the connections between nodes during training. Instead, it maintains a constant reservoir dynamic. The schematic of reservoir computing is shown in Figure 7.3. The use of reservoir computing enhances the efficiency of the training process, making it well-suited for real-time applications and hardware implementations. The reservoir functions as a computational framework, transforming input signals into multidimensional representations that include the fundamental structure of the data. Through the use of the reservoir's dynamics, complex calculations may be performed with high efficiency and resilience, even in the presence of noise and uncertainty. The compatibility of reservoir computing algorithms with hardware implementation is evident; nonetheless, the continuous issue lies in the development of efficient and reliable neuromorphic hardware [14–16].

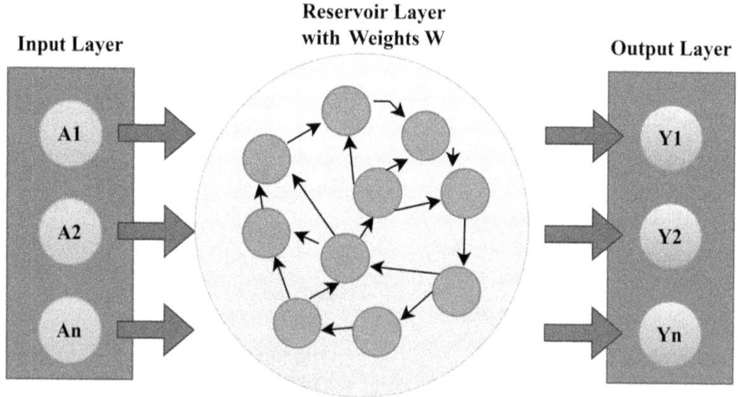

Figure 7.3 Schematic of reservoir computing.

7.3 APPLICATIONS OF NEUROMORPHIC COMPUTING

Various applications of neuromorphic computing are shown in Figure 7.4 and discussed in this section.

7.3.1 Neuromorphic sensory processing

Sensory processing systems represent a captivating application of neuromorphic computing, since they strive to emulate the operational capabilities of organic sensory organs, hence enhancing perception and interaction with the environment. The sensors include attributes such as minimal power consumption, a wide dynamic range, and event-driven functionality, making them well-suited for a variety of applications such as autonomous cars, surveillance systems, and medical apparatus. Neuromorphic sensors provide dependable and immediate perception in resource-constrained contexts by emulating the efficiency of organic sensory systems [17, 18].

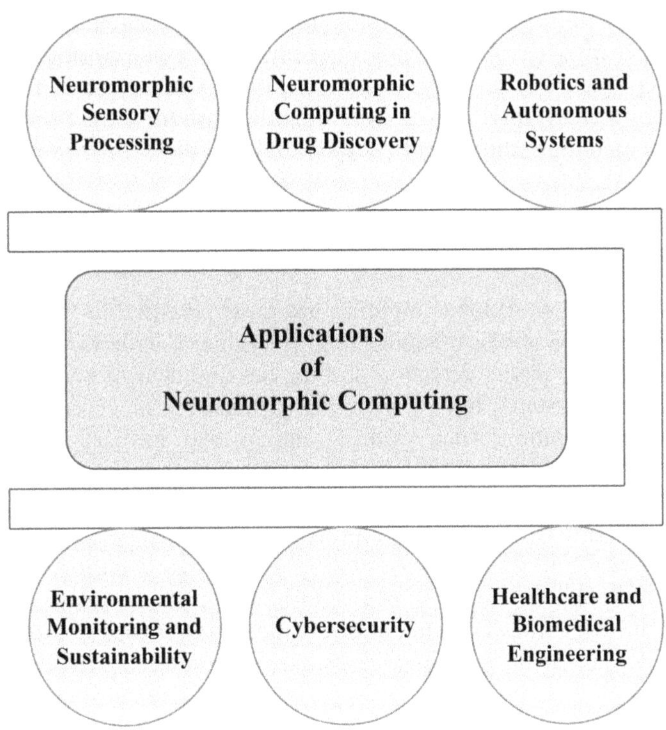

Figure 7.4 Applications of neuromorphic computing.

7.3.2 Neuromorphic computing in drug discovery

The pharmaceutical industry is progressively using neuromorphic computing as a means to accelerate the procedures involved in medication exploration and advancement. Neuromorphic systems facilitate the virtual screening of drugs, predict pharmacological efficacy, and enhance therapeutic target identification by simulating molecular interactions and biological processes. Simulated settings provide researchers with the ability to identify prospective drug candidates more effectively, resulting in reduced time and costs associated with traditional experimental methods. Moreover, neuromorphic models have the capacity to replicate the effects of pharmaceutical substances on brain circuits, therefore aiding in the development of therapeutic interventions for neurological disorders [19, 20].

7.3.3 Robotics and autonomous systems

Neuromorphic computing signifies a ground-breaking advancement in the field of robotics and autonomous systems. Neuromorphic processors enable robots to interact with their environment in real-time by emulating neural networks found in the human brain, hence replicating human cognitive capabilities. These systems have a notable capacity to adapt to dynamic settings, navigate complex terrains, and make prompt judgments with exceptional efficiency. Neuromorphic hardware integration accelerates the development of intelligent robots capable of performing a wide range of activities, including automated production and rescue missions in dangerous areas [21, 22].

7.3.4 Health care and biomedical engineering

The field of neuromorphic computing has great potential in the domains of health care and biomedical engineering, providing a wide range of applications that include illness detection and the development of brain prostheses. Neuromorphic systems have exceptional proficiency in processing delicate biological information, such as EEG signals and medical imaging, with enhanced accuracy and speed, by emulating the intricate neural networks of the brain. By enabling the continuous monitoring of physiological variables, early detection of abnormalities, and accurate delivery of therapeutic interventions in real-time, this technology sets the stage for customized health care. Moreover, the incorporation of neuromorphic implants exhibits potential in tackling the difficulties encountered by patients suffering from neurological illnesses, perhaps reinstating impaired sensory or motor capabilities [23, 24].

7.3.5 Cybersecurity

Neuromorphic computing greatly enhances cybersecurity defenses by efficiently detecting anomalies and implementing counterattack strategies.

Neuromorphic systems have the capacity to rapidly analyze large amounts of network traffic data, accurately detecting possible security breaches by emulating the brain's aptitude for pattern recognition and anomaly detection. These systems have the potential to adapt to constantly changing attack methods and learn from past security breaches, hence improving their effectiveness over time. Furthermore, the ability of neuromorphic technology to withstand hostile attacks makes it a crucial resource for strengthening essential infrastructure and protecting confidential information [25, 26].

7.3.6 Environmental monitoring and sustainability

The field of neuromorphic computing plays a crucial role in environmental monitoring and sustainability efforts by enabling the creation of intelligent sensor networks that improve the efficiency of data collection and processing. These networks effectively integrate neuromorphic sensors with edge computing platforms, enabling the continuous monitoring of diverse environmental variables such as air quality, temperature, and biodiversity in real time. Neuromorphic systems optimize energy consumption and bandwidth requirements by performing data processing at the edge and sending only relevant data. This makes them well-suited for deployment in distant and resource-constrained environments. In addition, these systems facilitate the process of predictive modeling and informed decision-making, thus contributing to the resolution of environmental challenges and the promotion of sustainable practices [27, 28].

7.4 CONCLUSION

Neuromorphic computing has the capability to execute specific computations with greater efficiency compared to conventional computing architectures, especially in activities related to pattern recognition, sensor data processing, and real-time decision-making. Neuromorphic computing seeks to imitate the parallelism, distributed nature, and energy efficiency of biological brain networks in contrast to standard computer designs that depend on binary logic gates and sequential processing. This chapter discussed the various computation methods and application areas of neuromorphic computing. The researchers are continuously investigating new hardware designs and algorithms to fully use the capabilities of neuromorphic systems and progress the area of intelligent computing.

REFERENCES

[1] Davies, Mike, Andreas Wild, Garrick Orchard, Yulia Sandamirskaya, Gabriel A. Fonseca Guerra, Prasad Joshi, Philipp Plank, and Sumedh R. Risbud. "Advancing neuromorphic computing with loihi: A survey of results and outlook." *Proceedings of the IEEE* 109, no. 5 (2021): 911–934.

[2] Park, Jaeyoung. "Neuromorphic computing using emerging synaptic devices: A retrospective summary and an outlook." *Electronics* 9, no. 9 (2020): 1414.

[3] Aimone, James B., Prasanna Date, Gabriel A. Fonseca-Guerra, Kathleen E. Hamilton, Kyle Henke, Bill Kay, Garrett T. Kenyon et al. "A review of non-cognitive applications for neuromorphic computing." *Neuromorphic Computing and Engineering* 2, no. 3 (2022): 032003.

[4] Nguyen, Duy-Anh, Xuan-Tu Tran, and Francesca Iacopi. "A review of algorithms and hardware implementations for spiking neural networks." *Journal of Low Power Electronics and Applications* 11, no. 2 (2021): 23.

[5] Taherkhani, Aboozar, Ammar Belatreche, Yuhua Li, Georgina Cosma, Liam P. Maguire, and T. Martin McGinnity. "A review of learning in biologically plausible spiking neural networks." *Neural Networks* 122 (2020): 253–272.

[6] Wang, Xiangwen, Xianghong Lin, and Xiaochao Dang. "Supervised learning in spiking neural networks: A review of algorithms and evaluations." *Neural Networks* 125 (2020): 258–280.

[7] Chatzidimitriou, Kyriakos C., and Pericles A. Mitkas. "Adaptive reservoir computing through evolution and learning." *Neurocomputing* 103 (2013): 198–209.

[8] Date, Prasanna, Bill Kay, Catherine Schuman, Robert Patton, and Thomas Potok. "Computational complexity of neuromorphic algorithms." In *International Conference on Neuromorphic Systems* 2021, pp. 1–7. 2021.

[9] Esser, Steven K., Paul A. Merolla, John V. Arthur, Andrew S. Cassidy, Rathinakumar Appuswamy, Alexander Andreopoulos, David J. Berg et al. "From the cover: Convolutional networks for fast, energy-efficient neuromorphic computing." *Proceedings of the National Academy of Sciences of the United States of America* 113, no. 41 (2016): 11441.

[10] Sun, Shengyang, Jiwei Li, Zhiwei Li, Haijun Liu, Qingjiang Li, and Xu Hui. "Low-consumption neuromorphic memristor architecture based on convolutional neural networks." In *2018 International Joint Conference on Neural Networks (IJCNN)*, pp. 1–6. IEEE, 2018.

[11] Huynh, Phu Khanh, M. Lakshmi Varshika, Ankita Paul, Murat Isik, Adarsha Balaji, and Anup Das. "Implementing spiking neural networks on neuromorphic architectures: A review." arXiv preprint arXiv:2202.08897 (2022).

[12] Davies, Mike, Narayan Srinivasa, Tsung-Han Lin, Gautham Chinya, Yongqiang Cao, Sri Harsha Choday, Georgios Dimou et al. "Loihi: A neuromorphic many-core processor with on-chip learning." *IEEE Micro* 38, no. 1 (2018): 82–99.

[13] Srinivasan, Gopalakrishnan, Chankyu Lee, Abhronil Sengupta, Priyadarshini Panda, Syed Shakib Sarwar, and Kaushik Roy. "Training deep spiking neural networks for energy-efficient neuromorphic computing." In *ICASSP 2020–2020 IEEE International Conference on Acoustics, Speech and Signal Processing (ICASSP)*, pp. 8549–8553. IEEE, 2020.

[14] Schrauwen, Benjamin, David Verstraeten, and Jan Van Campenhout. "An overview of reservoir computing: theory, applications and implementations." In *Proceedings of the 15th european symposium on artificial neural networks*, p. 471–482 2007.

[15] Cao, Jie, Xumeng Zhang, Hongfei Cheng, Jie Qiu, Xusheng Liu, Ming Wang, and Qi Liu. "Emerging dynamic memristors for neuromorphic reservoir computing." *Nanoscale* 14, no. 2 (2022): 289–298.

[16] Tanaka, Gouhei, Toshiyuki Yamane, Jean Benoit Héroux, Ryosho Nakane, Naoki Kanazawa, Seiji Takeda, Hidetoshi Numata, Daiju Nakano, and Akira Hirose. "Recent advances in physical reservoir computing: A review." *Neural Networks* 115 (2019): 100–123.
[17] Wan, Tianqing, Sijie Ma, Fuyou Liao, Lingwei Fan, and Yang Chai. "Neuromorphic sensory computing." *Science China Information* Sciences 65 (2022): 1–14.
[18] Liu, Shih-Chii, and Tobi Delbruck. "Neuromorphic sensory systems." *Current Opinion in Neurobiology* 20, no. 3 (2010): 288–295.
[19] Pastur-Romay, Lucas Antón, Francisco Cedrón, Alejandro Pazos, and Ana Belén Porto-Pazos. "Deep artificial neural networks and neuromorphic chips for big data analysis: pharmaceutical and bioinformatics applications." *International Journal of Molecular Sciences* 17, no. 8 (2016): 1313.
[20] Donati, Elisa, and Giacomo Indiveri. "Neuromorphic bioelectronic medicine for nervous system interfaces: from neural computational primitives to medical applications." *Progress in Biomedical Engineering* 5, no. 1 (2023): 013002.
[21] Piñero-Fuentes, Enrique, Salvador Canas-Moreno, Antonio Rios-Navarro, Tobi Delbruck, and Alejandro Linares-Barranco. "Autonomous driving of a rover-like robot using neuromorphic computing." In *International Work-Conference on Artificial Neural Networks*, pp. 57–68. Cham: Springer International Publishing, 2021.
[22] Aitsam, Muhammad, Sergio Davies, and Alessandro Di Nuovo. "Neuromorphic computing for interactive robotics: a systematic review." *IEEE Access* 10 (2022): 122261–122279.
[23] Azghadi, Mostafa Rahimi, Corey Lammie, Jason K. Eshraghian, Melika Payvand, Elisa Donati, Bernabe Linares-Barranco, and Giacomo Indiveri. "Hardware implementation of deep network accelerators towards healthcare and biomedical applications." *IEEE Transactions on Biomedical Circuits and Systems* 14, no. 6 (2020): 1138–1159.
[24] Tian, Fengshi, Jie Yang, Shiqi Zhao, and Mohamad Sawan. "NeuroCARE: A generic neuromorphic edge computing framework for healthcare applications." *Frontiers in Neuroscience* 17 (2023): 1093865.
[25] Zahm, Wyler, Tyler Stern, Malyaban Bal, Abhronil Sengupta, Aswin Jose, Suhas Chelian, and Srini Vasan. "Cyber-neuro RT: Real-time neuromorphic cybersecurity. *Procedia Computer Science* 213 (2022): 536–545.
[26] Pino, Robinson E., and Alexander Kott. "Neuromorphic computing for cognitive augmentation in cyber defense." *Cybersecurity Systems for Human Cognition Augmentation* 61, (2014): 19–45.
[27] Vogginger, Bernhard, Amirhossein Rostami, Vaibhav Jain, Sirine Arfa, Andreas Hantsch, David Kappel, Michael Schäfer et al. "Neuromorphic hardware for sustainable AI data centers." arXiv preprint arXiv:2402.02521 (2024).
[28] Fan, Zhencheng, Zheng Yan, and Shiping Wen. "Deep learning and artificial intelligence in sustainability: a review of SDGs, renewable energy, and environmental health." *Sustainability* 15, no. 18 (2023): 13493.

Index

Pages in *italics* refer to figures and pages in **bold** refer to tables.

For Product Safety Concerns and Information please contact our EU representative GPSR@taylorandfrancis.com
Taylor & Francis Verlag GmbH, Kaufingerstraße 24, 80331 München, Germany

www.ingramcontent.com/pod-product-compliance
Ingram Content Group UK Ltd.
Pitfield, Milton Keynes, MK11 3LW, UK
UKHW022204310726
473902UK00009B/525
* 9 7 8 1 0 3 2 9 6 8 7 7 3 *